T0198354

MACHINES

BEHAVING

BADLY

MACHINES BEHAVING BADLY

THE MORALITY OF AI

TOBY WALSH

Published 2022 by arrangement with Black Inc.
First published in Australia and New Zealand by La Trobe University Press, 2022

FLINT is an imprint of The History Press
97 St George's Place, Cheltenham,
Gloucestershire, GL50 3QB
www.flintbooks.co.uk

British Library Cataloguing in Publication Data.
A catalogue record for this book is available from the British Library.

ISBN 978 0 7509 9936 6

Cover design by Tristan Main
Text design and typesetting by Tristan Main
Cover illustrations by Lemonsoup14 / Shutterstock, Mykola / Adobe Stock
Printed and bound in Great Britain by TJ Books Limited, Padstow, Cornwall.

Trees for LYfe

To A and B,
my A to Z

But remember, please, the Law by which we live,
    We are not built to comprehend a lie,
We can neither love nor pity nor forgive.
    If you make a slip in handling us you die!
We are greater than the Peoples or the Kings—
    Be humble, as you crawl beneath our rods!—
Our touch can alter all created things,
    We are everything on earth—except The Gods!

Though our smoke may hide the Heavens from your eyes,
It will vanish and the stars will shine again,
Because, for all our power and weight and size,
We are nothing more than children of your brain!

FROM 'THE SECRET OF THE MACHINES',
BY RUDYARD KIPLING

CONTENTS

AI

You surely know what artificial intelligence is. After all, Hollywood has given you plenty of examples.

Artificial intelligence is the terrifying T-800 robot played by Arnold Schwarzenegger in the *Terminator* movies. It is Ava, the female humanoid robot in *Ex Machina* that deceives humans to enable it to escape from captivity. It is the Tyrell Corporation Nexus-6 replicant robot in *Blade Runner*, trying to save itself from being 'retired' by Harrison Ford.

My personal favourite is HAL 9000, the sentient computer in *2001: A Space Odyssey*. HAL talks, plays chess, runs the space station – and has murderous intent. HAL voices one of the most famous lines ever said by a computer: 'I'm sorry, Dave. I'm afraid I can't do that.'

Why is it that the AI is always trying to kill us?

In reality, artificial intelligence is none of these conscious robots. We cannot yet build machines that match the intelligence of a two-year-old. We can, however, program computers to do narrow, focused tasks that humans need some sort of intelligence to solve. And that has profound consequences.

If artificial intelligence is not the stuff of Hollywood movies, then what is it? Oddly enough, AI is already part of our lives. However, much of it is somewhat hidden from sight.

Every time you ask Siri a question, you are using artificial intelligence. It is speech recognition software that converts your speech into a natural language question. Then natural language processing algorithms convert this question into a search query. Then search algorithms answer this query. And then ranking algorithms predict the most 'useful' search results.

If you're lucky enough to own a Tesla, you can sit in the driver's seat, not driving, while the car drives itself autonomously along the highway. It uses a whole host of AI algorithms that sense the road and environment, plan a course of action and drive the car to where you want to go. The AI is smart enough that, in these limited circumstances, you can trust it with your life.

Artificial intelligence is also the machine-learning algorithms that predict which criminals will reoffend, who will default on their loans, and whom to shortlist for a job. AI is touching everything from the start of life, predicting which fertilised eggs to implant, to the very end, powering chatbots that spookily bring back those who have died.

For those of us working in the field, the fact that AI often falls out of sight in this way is gratifying evidence of its success. Ultimately, AI will be a pervasive and critical technology, like electricity, that invisibly permeates all aspects of our lives.

Almost every device today uses electricity. It is an essential and largely unseen component of our homes, our cars, our farms, our factories and our shops. It brings energy and data to almost everything we do. If electricity disappeared, the world would quickly grind to a halt. In a similar way, AI will shortly become an indispensable and mostly invisible component of our lives. It is already providing the smartness in our smartphones. And soon it will be powering the intelligence in our self-flying cars, smart cities, and intelligent offices and factories.

A common misconception is that AI is a single thing. Just like our intelligence is a collection of different skills, AI today is a collection of different technologies, such as machine learning, natural language processing and speech recognition. Because many of the recent advances in AI have been in the area of machine learning, artificial intelligence is often mistakenly conflated with it. However, just as humans do more than simply learn how to solve tasks, AI is about more than just machine learning.

We are almost certainly at the peak of inflated expectations in the hype cycle around AI. And we will likely descend shortly into a trough of disillusionment as reality fails to match expectations. If you added up everything written in the newspapers about the progress being made, or believed the many optimistic surveys, you might suspect that computers will soon be matching or even surpassing humans in intelligence.

The reality is that while we have made good progress in getting machines to solve narrow problems, we have made almost no progress on building more general intelligence that can tackle a wide range of problems. Nevertheless, it is impossible to list all the narrow applications that AI is now being used in, but I will mention a few in order to illustrate the wide variety. AI is currently being used to:

- detect malware

- predict hospital admissions

- check legal contracts for errors

- prevent money laundering

- identify birds from their song

- predict gene function

- discover new materials

- mark essays

- identify the best crops to plant, and

- (controversially) predict crime and schedule police patrols.

Indeed, you might think it would be easier to list the areas where AI is *not* being used – except that it's almost impossible to think of any such area. Anyway, what this makes clear is that AI shows significant promise for transforming our society.

The potential advantages of AI encompass almost every sector, and include agriculture, banking, construction, defence, education, entertainment, finance, government, healthcare, housing, insurance, justice, law, manufacturing, mining, politics, retail and transportation.

The benefits of AI are not purely economic. Artificial intelligence also offers many opportunities for us to improve our societal and environmental wellbeing. It can, for example, be used to make buildings and transportation more efficient, help conserve the planet's limited resources, provide vision to those who cannot see, and tackle many of the wicked problems facing the world, like the climate emergency.

Alongside these benefits, AI also presents significant risks. These include the displacement of jobs, an increase in inequality within and between countries, the transformation of war, the corrosion of political discourse, and the erosion of privacy and other human rights. Indeed, we are already seeing worrying trends in many of these areas.

STRANGE INTRUDERS

One of the challenges of any new technology are the unexpected consequences. As the social critic Neil Postman put it in 1992, we 'gaze on technology as a lover does on his beloved, seeing it as without blemish and entertaining no apprehension for the future'.[1] Artificial intelligence is no exception. Many – and I count myself among them – look lovingly upon its immense potential. It has been called by some our 'final invention'. And the unexpected consequences of AI may be the most consequential of any in human history.

In a 1998 speech titled 'Five Things We Need to Know about Technological Change', Postman summarised many of the issues that should concern you today about AI as it takes on ever more important roles in your life.[2] His words ring even truer now than they did almost 25 years ago. His first advice:

Technology giveth and technology taketh away. This means that for every advantage a new technology offers, there is always a corresponding disadvantage. The disadvantage may exceed in importance the advantage, or the advantage may well be worth the cost ... the advantages and disadvantages of new technologies are never distributed evenly among the population. This means that every new technology benefits some and harms others.

He warned:

That is why we must be cautious about technological innovation. The consequences of technological change are always vast, often unpredictable and largely irreversible. That is also why we must be suspicious of capitalists. Capitalists are by definition not only personal risk takers but, more to the point, cultural risk takers. The most creative and daring of them hope to exploit new technologies to the fullest, and do not much care what traditions are overthrown in the process or whether or not a culture is prepared to function without such traditions. Capitalists are, in a word, radicals.

And he offered a suggestion:

The best way to view technology is as a strange intruder, to remember that technology is not part of God's plan but a product of human creativity and hubris, and that its capacity for good or evil rests entirely on human awareness of what it does for us and to us.

He concluded his speech with a recommendation:

> In the past, we experienced technological change in the manner of sleep-walkers. Our unspoken slogan has been 'technology über alles', and we have been willing to shape our lives to fit the requirements of technology, not the requirements of culture. This is a form of stupidity, especially in an age of vast technological change. We need to proceed with our eyes wide open so that we may use technology rather than be used by it.

The goal of this book is to open your eyes to this strange intruder, to get you to think about the unintended consequences of AI.

History provides us with plenty of troubling examples of the unintended consequences of new technologies. When Thomas Savery patented the first steam-powered pump in 1698, no one was worrying about global warming. Steam engines powered the Industrial Revolution, which ultimately lifted millions out of poverty. But we are now seeing the unintended consequences of all that the steam engine begat today, both literally and metaphorically. The climate is changing, and millions are starting to suffer.

In 1969, when the first Boeing 747 took to the air, the age of affordable air travel began. It seems to have been largely forgotten, but the world at that time was in the midst of a deadly pandemic. This was caused by a strain of the influenza virus known as 'the Hong Kong flu'. It would kill over a million people. No one, however, was concerned that the 747 was going to make things worse. But by making the world smaller, the 747 almost certainly made the current COVID-19 global pandemic much deadlier.

Can we ever hope, then, to predict the unintended consequences of AI?

WARNING SIGNS

Artificial intelligence offers immense potential to improve our wellbeing, but equally AI could be detrimental to the planet. So far, we have been very poor at heeding any warning signs. Let me give just one example.

In 1959, a data science firm called the Simulmatics Corporation was founded, with the goal of using algorithms and large data sets to target voters and consumers. The company's first mission was to win back the White House for the Democratic Party and install John F. Kennedy as president. The company used election returns and public-opinion surveys going back to 1952 to construct a vast database that sorted voters into 480 different categories. The company then built a computer simulation of the 1960 election in which they tested how voters would respond to candidates taking different positions.

The simulations highlighted the need to win the Black vote, and that required taking a strong position on civil rights. When Martin Luther King Jr was arrested in the middle of the campaign, JFK famously called King's wife to reassure her, while his brother, Robert F. Kennedy, called a judge the next day to help secure King's release. These actions undoubtably helped the Democratic candidate win many Black votes.

The computer simulations also revealed that JFK needed to address the issue of his Catholicism and the prevailing prejudices against this. JFK followed this advice and talked openly about his religious beliefs. He would become the first (and, until Joe Biden, the only) Catholic president of the United States.

On the back of this success, Simulmatics went public in 1961, promising investors it would 'engage principally in estimating probable human behavior by the use of computer technology'. This was a disturbing promise. By 1970 the company was bankrupt; it would remain largely forgotten until quite recently.[3]

You've probably noticed that the story of Simulmatics sounds eerily similar to that of Cambridge Analytica before its own bankruptcy in 2018. Here was another company mining human data to manipulate US elections. Perhaps more disturbing still is that this problem had been predicted at the very dawn of computing, by Norbert Wiener in his classic and influential text *The Human Use of Human Beings: Cybernetics and Society*.[4]

Wiener saw past the optimism of Alan Turing and others to identify a real danger posed by the recently invented computer. In the penultimate chapter of his book, he writes:

> [M]achines ... may be used by a human being or a block of human beings to increase their control over the rest of the race or that political leaders may attempt to control their populations by means not of machines themselves but through political techniques as narrow and indifferent to human possibility as if they had, in fact, been conceived mechanically.

The chapter then ends with a warning: 'The hour is very late, and the choice of good and evil knocks at our door.'

Despite these warnings, we walked straight into this political minefield in 2016, first with the Brexit referendum in the United Kingdom and then with the election of Donald Trump in the United States. Machines are now routinely treating humans mechanically and controlling populations politically. Wiener's prophecies have come true.

BREAKING BAD

It's not as if the technology companies have been hiding their intentions. Let's return to the Cambridge Analytica scandal. Much of the public concern was about how Facebook helped Cambridge Analytica harvest people's private information without their consent. And this was, of course, bad behaviour all round.

But there's a less discussed side to the Cambridge Analytica story, which is that this stolen information was then used to manipulate how people vote. In fact, Facebook had employees working full-time in the Cambridge Analytica offices in Tucson, Arizona, helping it micro-target political adverts. Cambridge Analytica was one of Facebook's best customers during the 2016 elections.[5]

It's hard to understand, then, why Facebook CEO Mark Zuckerberg sounded so surprised when he testified to Congress in April 2018 about what had happened.[6] Facebook had been a very active player in manipulating the vote. And manipulating voters has been bad behaviour for thousands of years, ever since the ancient Greeks. We don't need any new ethics to decide this.

What's worse is that Facebook had been doing this for many years. Facebook published case studies from as far back as 2010 describing elections where they had been actively changing the outcome. They boasted that 'using Facebook as a market research tool and as a platform for ad saturation can be used to change public opinion in any political campaign'.

You can't be clearer than this. Facebook can be used to change public opinion in *any* political campaign. These damaging claims remain online on Facebook's official Government, Politics and Advocacy pages today.[7]

These examples highlight a fundamental ethical problem, a dangerous truth somewhat overlooked by advertisers and political pollsters. Human minds can be easily hacked. And AI tools like machine learning put this problem on steroids. We can collect data on a population and change people's views at scale and at speed, and for very little cost.

When this sort of thing was done to sell washing powder, it didn't matter so much. We were always going to buy some washing powder, and whether advertising persuaded us to buy OMO or Daz wasn't really a big deal. But now it's being done to determine who becomes president of the United States. Or whether Britain exits the European Union. It matters a great deal.

This book sets out to explore these and other ethical problems which artificial intelligence is posing. It asks many questions. Can we build machines that behave ethically? What other ethical challenges does AI create? And what lies in store for humanity as we build ever more amazing and intelligent machines?

THE PEOPLE

THE GEEKS TAKING OVER

To understand why ethical concerns around artificial intelligence are rampant today, it may help to know a little about the people who are building AI. It is perhaps not widely recognised how small this group actually is. The number of people with a PhD in AI – making them the people who truly understand this rather complex technology – is measured in the tens of thousands.[1] There may never have been a planet-wide revolution before which was driven by such a small pool of people.

What this small group is building is partly a reflection of who they are. And this group is far from representative of the wider society in which that AI is being used. This has created, and will continue to create, fundamental problems, many of which are of an ethical nature.

Let me begin with an observation. It's a rather uncomfortable one for someone who has devoted his adult life to trying to build artificial intelligence, and who spent much of his childhood dreaming of it too. There's no easy way to put this. The field of AI attracts some odd people. And I should probably count myself as one of them.

Back in pre-pandemic times, AI researchers like me would fly to the farthest corners of the world. I never understood how a round Earth could have 'farthest corners' … Did we inherit them from flat Earth times? Anyway, we would go to conferences in these faraway places to hear about the latest advances in the field.[2] AI is studied and developed on all the continents of the globe, and as a consequence AI conferences are also held just about everywhere you can think.[3]

On many of these trips, my wife would sit next to me at an airport and point out one of my colleagues in the distance. 'That must be one of yours,' she would say, indicating a geeky-looking person. She was invariably correct: the distinctive person in the distance would be one of my colleagues.

But the oddness of AI researchers is more than skin-deep. There's a particular mindset held by those in the field. In artificial intelligence, we build models of the world. These models are much simpler and better behaved than the real one. And we become masters of these artificial universes. We get to control the inputs and the outputs. And everything in between. The computer does precisely and only what we tell it to do.

The day I began building artificial models like this, more than 30 years ago, I was seduced. I remember well my first AI program: it found proofs of simple mathematical statements. It was written in an exotic programming language called Prolog, which was favoured by AI researchers at that time.

I gave my AI program the task of proving a theorem that, I imagined, was well beyond its capability. There are some beautiful theorems by Alan Turing, Kurt Gödel and others that show that no computer program, however complex and sophisticated, can prove all mathematical statements. But my AI program didn't come close to testing these fundamental limits.

I asked my program to prove a simple mathematical statement: the Law of the Excluded Middle. This is the law that every proposition is either true or false. In symbols, 'P or not P'. Either $2^{82,589,933} -1$ is prime or it isn't.[4] Either the stock market will crash next year or it won't. Either the Moon is made of cheese or it isn't. This is a mathematical truth that can be traced back through Leibnitz to Aristotle, over two millennia ago.

I almost fell off my chair in amazement when my AI program spat out a proof. It is not the most complex proof ever found by a computer program, by a long margin. But this is a proof that defeats many undergraduate students who are learning logic for the first time. And I was the creator of this program. A program that was the master of this mathematical universe. Admittedly, it was a very simple universe – but thoughts about mastering even a simple universe are dangerous.

The real world doesn't bend to the simple rules of our artificial universes. We're a long way from having computer programs that can take over many facets of human decision-making. Indeed, it's not at all clear if computers will ever match humans in all their abilities: their cognitive, emotional and social intelligence, their creativity, and their adaptability. But the field of AI is full of people who would like life to be a simple artificial universe that our computers could solve. And for many years I was one of them.

THE SEA OF DUDES

One especially problematic feature of the group building these artificial universes has been dubbed the 'sea of dudes' problem. This phrase was coined in 2016 by Margaret Mitchell, then an AI researcher at Microsoft Research and who, in 2021, was fired from Google in controversial circumstances. The phrase highlights the fact that very few AI researchers are women.

Stanford's AI index, which tracks progress in AI, reports that the number of women graduating with a PhD in AI in the United States has remained stable at around 20 per cent for the last decade. The figure is similar in many other countries, and the numbers are not much better at the undergraduate level. This is despite many ongoing efforts to increase diversity.

Actually, Margaret Mitchell might have more accurately described it as a 'sea of white dudes' problem. Not only are four-fifths of AI researchers male, they are also mostly white males.[5] Black, Hispanic and other groups are poorly represented within AI, both in academia and in industry.

There is little data on the extent of the racial problem in AI, which itself is a problem. However, it is a very *visible* problem. Timnit Gebru is an AI and ethics researcher who was fired in controversial circumstances by Google in late 2020. As a PhD student, she co-founded Black in AI after counting just six Black AI researchers out of the 8500 researchers attending NIPS, the largest AI conference in 2016.

Even the name of that conference, NIPS, hints at the issues. In 2018, the NIPS conference rebranded itself NeurIPS to distance itself from the sexist and racial associations of its previous acronym. Other nails in the coffin of the conference's old acronym included the 2017 pre-conference's 'counter-culture' event, TITS, along with the conference T-shirts carrying the dreadful slogan 'My NIPS are NP-hard'. To understand this geeky joke, you have to know that 'NP-hard' is a technical term for a computationally challenging problem. But it doesn't take a geeky background to understand the sexism of the slogan.

Anima Anandkumar, a California Institute of Technology (Caltech) professor and director of machine-learning research at Nvidia, led the #ProtestNIPS campaign. Sadly, she reported that she was trolled and harassed on social media by a number of senior male AI researchers for calling for change. Nevertheless, pleasingly and appropriately, the name change went ahead.

Racial, gender and other imbalances are undoubtably harmful to progress in developing AI, especially in ensuring that AI does not disadvantage some of these groups. There will be questions not asked and problems not addressed because of the lack of diversity in the room. There is plentiful evidence that diverse groups build better product. Let me give two simple examples to illustrate this claim.

When the Apple Watch was first released in 2015, the application programming interface (API) used to build health apps didn't track any aspect of a woman's menstrual cycle. The mostly male Apple developers appear not to have thought it important enough to include. Yet you cannot properly understand a woman's health without taking account of her menstrual cycle. Since 2019, the API has corrected this oversight.

A second example: Joy Buolamwini, an AI researcher at the Massachusetts Institute of Technology (MIT) has uncovered serious racial and gender biases in the facial-recognition software being used by companies such as Amazon and IBM. This software frequently fails to identify the faces of people from disadvantaged groups, especially those of darker-skinned women. Buolamwini eventually had to resort to wearing a white mask for the face-detecting software to detect her face.

THE GODFATHERS OF AI

Alongside the 'sea of dudes', another problem is the phrase 'the godfathers of AI'. This refers to Yoshua Bengio, Geoffrey Hinton and Yann LeCun, a famous trio of machine-learning researchers who won the 2018 Turing Award, the Nobel Prize of computing, for their pioneering research in the subfield of deep learning.

There is much wrong with the idea that Bengio, Hinton and LeCun are the 'godfathers' of AI. First, it supposes that AI is just deep learning. This ignores all the other successful ideas developed in AI that are already transforming your life.

The next time you use Google Maps, for instance, please pause to thank Peter Hart, Nils Nilsson and Bertram Raphael for their 1968 route-finding algorithm.[6] This algorithm was originally used to direct Shakey, the first fully autonomous robot, who, as the name suggests, tended to shake a little too much. It has since been repurposed to guide us humans around a map. It's somewhat ironic that one of the most common uses of AI today is to guide not robots but humans. Alan Turing would doubtless be amused.

And the next time you read an email, please pause to thank the Reverend Thomas Bayes. Back in the seventeenth century, Bayes discovered what is now known as Bayes' rule for statistical inference. Bayes' rule has found numerous applications in machine learning, from spam filters to detecting nuclear weapons tests. Without the Reverend's insights, you would be drowning in junk emails.

We should also not forget the many other people outside of deep learning who laid the intellectual foundations of the field of artificial intelligence. This list starts with Alan Turing, whom *Time* named as one of the 100 most important people of the twentieth century.[7] In 1000 years' time, if the human race has not caused its own extinction, I suspect Turing might be considered the most important person of the twentieth century. He was a founder not just of the field of artificial intelligence but of the whole of computing. If there is one person who should be called a godfather of AI, it is Alan Turing.

But even if you limit yourself to deep learning, which admittedly has had some spectacular successes in recent years, there are many other people who deserve credit. Back propagation, the core algorithm used to update weights in deep learning, was popularised by just one of this trio, Geoffrey Hinton. However, it was based on work he did with David Rumelhart and Ronald Williams in the late 1980s.[8] Many others also deserve credit for back propagation, including Henry Kelley in 1960, Arthur Bryson in 1961, Stuart Dreyfus in 1962 and Paul Werbos in 1974.

Even this ignores many other people who made important intellectual contributions to deep learning, including Jürgen Schmidhuber, who developed Long Short-Term Memory (LSTM), which is at the heart of many deep networks doing speech recognition, and is used in Apple's Siri, Amazon's Alexa and Google's Voice Search; my friend Rina Dechter, who actually coined the term 'deep learning'; Andrew Ng, who imaginatively repurposed GPUs from processing graphics to tackle the computational challenge of training large deep networks;[9] and Fei-Fei Li, who was behind ImageNet, the large data set of images that has driven many advances in this area.

Putting aside all these academic concerns, there remains a fundamental problem with the term 'godfathers of AI'. It supposes artificial intelligence has godfathers and not godmothers. This slights the many women who have made important contributions to the field, including:

- Ada Lovelace, the *first* computer programmer ever and someone who, back in the eighteenth century, pondered whether computers would be creative

- Kathleen McNulty, Frances Bilas, Betty Jean Jennings, Ruth Lichterman, Elizabeth Snyder and Marlyn Wescoff, who were originally human 'computers', but went on to be the programming team of ENIAC, the first electronic general-purpose digital computer

- Grace Hopper, who invented one of the first high-level programming languages and discovered the first ever computer bug[10]

- Karen Spärck Jones, who did pioneering work in natural language processing that helped build the modern search engine, and

- Margaret Boden, who developed the world's first academic
  program in cognitive science, and explored the ideas on
  AI and creativity first discussed by Ada Lovelace.

The term 'godfathers of AI' also disregards the many women, young and old, who are making important contributions to AI today. This includes amazing researchers like Cynthia Breazeal, Carla Brodley, Joy Buolamwini, Diane Cook, Corinna Cortes, Kate Crawford, Rina Dechter, Marie desJardins, Edith Elkind, Timnit Gebru, Lise Getoor, Yolanda Gil, Maria Gini, Carla Gomes, Kristen Grauman, Barbara Grosz, Barbara Hayes-Roth, Marti Hearst, Leslie Kaelbling, Daphne Koller, Sarit Kraus, Fei-Fei Li, Deborah McGuinness, Sheila McIlraith, Pattie Maes, Maja Matarić, Joelle Pineau, Martha Pollack, Doina Precup, Pearl Pu, Daniela Rus, Cordelia Schmid, Dawn Song, Katia Sycara, Manuela Veloso and Meredith Whittaker, to name just a few.

I very much hope, therefore, that we follow Trotsky's advice and consign the phrase 'godfathers of AI' to the dustbin of history.[11] If we need to talk about the people responsible for some of the early breakthroughs, there are better phrases, such as the 'AI pioneers'.

THE CRAZY VALLEY

Artificial intelligence is, of course, being built around the world. I have friends working on AI everywhere, from Adelaide to Zimbabwe. But one special hothouse is Silicon Valley. The Valley is close to Stanford University, where the late John McCarthy, the person who named the field, set up shop in the 1960s and laid many of the foundation stones of AI.[12]

The Valley is home to the largest concentration of venture capitalists on the planet. The United States is responsible for about two-thirds of all venture capital funding worldwide, and half of this goes into the Valley. In other words, venture capital funding can be broken into three roughly equally sized parts: Silicon Valley (which has now spread out into the larger Bay Area), the rest of America, and the rest of the world. Each of these three parts is worth around $25 billion per year.[13] To put that into perspective, each slice of this venture capital pie is about equal to the gross domestic product of a small European nation like Estonia.

This concentration of venture funding has meant that much of the AI that has entered our lives came out of Silicon Valley. And much of that has been funded by a small number of venture capital firms based on Sand Hill Road. This unassuming road runs through Palo Alto and Menlo Park in Silicon Valley. Real estate here is more expensive than almost anywhere else in the world, often exceeding that in Manhattan or London's West End.

Many of the most successful venture capital firms on the planet are based on Sand Hill Road, including Andreessen Horowitz and Kleiner Perkins. Andreessen Horowitz was an early investor in well-known companies such as Facebook, Groupon, Airbnb, Foursquare and Stripe, while Kleiner Perkins was an early investor in Amazon, Google, Netscape and Twitter.

Anyone who has spent time in the Valley knows it is a very odd place. The coffee shops are full of optimistic 20-year-olds with laptops, working for no pay, discussing their plans to create ventures such as the 'Uber for dog walking'. They hope to touch the lives of billions. Given that there are estimated to be fewer than 200 million pet dogs on the planet, it's not clear to me how Uber Dogs will touch a billion people, but that's not stopping them.[14]

I often joke that there's a strange Kool Aid that folks in the Valley drink. But it really seems that way. The ethos of the place is that you haven't succeeded if you haven't failed. Entrepreneurs wear their failures with pride – these experiences have primed them, they'll tell you, for success the next time around.

And there have been some spectacular failures. Dotcom flops like Webvan, which burnt through half a billion dollars. Or the UK clothing retailer boo.com, which spent $135 million in just 18 months before going bankrupt. Or Munchery, a food delivery website that you've probably never heard of before today – it went through over $100 million before closing shop.

No idea seems too stupid to fund. Guess which one of the following companies I made up. The company with a messaging app that lets you send just one word: 'Yo.' The company that charges you $27 every month to send you $20 in quarters, so you'll have change to do your washing. The company that sends you natural snow in the post. Or the company building a mind-reading headset for your dog, which doesn't actually work.

Okay, I'll admit it – I was messing with you. None of these companies was made up. All were funded by venture capital. And, not surprisingly, all eventually went bust.

THE SHADOW OF AYN RAND

A long shadow cast over many in the Valley is that of one of its darlings, the philosopher Ayn Rand. Her novel *Atlas Shrugged* was on the *New York Times'* Bestseller List for 21 weeks after its publication in 1957. And sales of her book have increased almost every year since, annually hitting over 300,000 copies. In 1991, the Book of the Month Club and the Library of Congress asked readers to name the most influential book in their lives. *Atlas Shrugged* came in second. First place went to the Bible.

Many readers of *Atlas Shrugged*, especially in the tech community, relate to the philosophy described in this cult dystopian book. Rand called this philosophy 'objectivism'. It rejected most previous philosophical ideas in favour of the single-minded pursuit of individual happiness. Somewhat immodestly, given the rich and long history of philosophy, the author would only recommend the three As: Aristotle, Aquinas and Ayn Rand.[15] From Aristotle, she borrowed an emphasis on logical reasoning. While she rejected all religion on the grounds of its conflicts with rationality, she recognised Thomas Aquinas for lightening the Dark Ages by his promotion of the works of Aristotle. And from her own life, she focused on the struggle between the individual and the state that played out from her birth in Saint Petersburg, her emigration from Russia aged 21, to the naked capitalism of New York City, where she settled.

Rand's objectivism elevated rational thought above all else. According to her, our moral purpose is to follow our individual self-interest. We have direct contact with reality via our perception of the world. And we gain knowledge with which to seek out this happiness either from such perception or by reasoning about what we learn from such perception.

Rand considered the only social system consistent with objectivism to be laissez-faire capitalism. She opposed all other social systems, be they socialism, monarchism, fascism or, unsurprisingly given her birthplace, communism. For Rand, the role of the state was to protect individual rights so that individuals could go about their moral duty of pursuing happiness. Predictably, then, many libertarians have found considerable comfort in *Atlas Shrugged*.

But objectivism doesn't just provide a guide to live one's life. It could also be viewed as an instruction book for building an artificial intelligence. Rand wrote in *Atlas Shrugged*:

> Man's mind is his basic tool of survival … To remain alive he must act and before he can act he must know the nature and purpose of his action. He cannot obtain his food without knowledge of food and of the way to obtain it. He cannot dig a ditch – or build a cyclotron – without a knowledge of his aim and the means to achieve it. To remain alive, he must think.

Putting aside the quotation's dated sexism, much the same could be said of an artificial intelligence. The basic tool of survival for a robot is its ability to reason about the world. A robot has direct contact with the reality of the world via its perception of that world. Its sole purpose is to maximise its reward function. And it does so by reasoning rationally about those precepts.

It is perhaps unsurprising, then, that Rand's work appeals to many AI researchers. She laid out a philosophy that described not just their lives, but the inner workings of the machines they are trying to build. What could be more seductive? As a consequence, Ayn Rand has become Silicon Valley's favourite philosopher queen. And many start-ups and children in the Valley are named after the people and institutions in her books.

TECHNO-LIBERTARIANS

Moving beyond objectivism, we come to libertarianism, and that special form of libertarianism found in Silicon Valley, techno-libertarianism. This philosophical movement grew out of the hacker culture that emerged in places like the AI Lab at MIT, and other AI hotbeds like the computer science departments at Carnegie Mellon University and the University of California at Berkeley.

Techno-libertarians wish to minimise regulation, censorship and anything else that gets in the way of a 'free' technological future. Here, 'free' means without restrictions, not without cost. The best solution for a techno-libertarian is a free market built with some fancy new technology like the blockchain, where behaving rationally is every individual's best course of action.

John Perry Barlow's 1996 *Declaration of the Independence of Cyberspace* is perhaps the seminal techno-libertarian creed.[16] Let me quote a few paragraphs from both the beginning and the end of the *Declaration* that capture its spirit:

Governments of the Industrial World, you weary giants of flesh and steel, I come from Cyberspace, the new home of Mind. On behalf of the future, I ask you of the past to leave us alone.

You are not welcome among us. You have no sovereignty where we gather.

We have no elected government, nor are we likely to have one, so I address you with no greater authority than that with which liberty itself always speaks. I declare the global social space we are building to be naturally independent of the tyrannies you seek to impose on us. You have no moral right to rule us nor do you possess any methods of enforcement we have true reason to fear ...

Cyberspace consists of transactions, relationships, and thought itself, arrayed like a standing wave in the web of our communications. Ours is a world that is both everywhere and nowhere, but it is not where bodies live ...

In our world, whatever the human mind may create can be reproduced and distributed infinitely at no cost. The global conveyance of thought no longer requires your factories to accomplish.

These increasingly hostile and colonial measures place us in the same position as those previous lovers of freedom and self-determination who had to reject the authorities of distant, uninformed powers. We must declare our virtual selves immune to your sovereignty, even as we continue to consent to your rule over our bodies. We will spread ourselves across the Planet so that no one can arrest our thoughts.

We will create a civilization of the Mind in Cyberspace. May it be more humane and fair than the world your governments have made before.

Not surprisingly, therefore, artificial intelligence is immensely attractive to techno-libertarians. But alongside the 'Mind in Cyberspace' that AI may help build, we have to contend with all the other baggage of techno-libertarianism. And some of this is already proving unattractive.

Techno-libertarians believe, for example, that you can't and you shouldn't regulate cyberspace. You can't because digital bits aren't physical. And tech companies span national boundaries, so they cannot be bound by national rules. And you shouldn't regulate cyberspace because, even if you could, regulation will only stifle innovation.

Fortunately, many politicians are starting to realise that you can and indeed should regulate cyberspace. You can regulate it because digital bits live on physical servers, and these are located in particular national jurisdictions. Additionally, tech companies act in national markets where rules apply. And you should regulate cyberspace because every market needs rules to ensure it operates efficiently and in consumers' best interests. We've had to regulate every other large sector before: banks, pharmaceutical companies, telecommunication firms and oil companies, to name just a few. Now that the tech industry is one of the largest sectors of the economy, it is overdue some proper regulation for the public good.

TRANSHUMANISTS

Alongside objectivism and techno-libertarianism, another shadow cast over artificial intelligence is the shadow of transhumanism.

This is the dream of transcending the physical limitations of our imperfect bodies. For many transhumanists, this extends even into a dream of living forever. The field of AI includes a surprising number of transhumanists, such as Ray Kurzweil, Nick Bostrom and the Turing Award–winner Marvin Minsky.

Indeed, Minsky was one of the founders of the field of artificial intelligence. He was one of the organisers of the Dartmouth conference in 1956, the event that kicked off the field and at which the term 'artificial intelligence' was created.

Unlike Kurzweil and Bostrom, Minksy is no longer alive. He died in January 2016. And his body is preserved at –195.79 degrees Celsius, in a bath of liquid nitrogen, at the Alcor Life Extension foundation in Scottsdale, Arizona, awaiting advances in medical science that might bring him back to life. In due course, Kurzweil and Bostrom will be joining Minsky in Scottsdale for their own suspended animation.

Artificial intelligence offers both a promise and a danger for transhumanists. The promise of AI is that one day we might be able to upload our brains into a computer. We can thereby escape the limitations of our biology and gain digital immortality. Unlike our physical selves, digitals bits need never decay. However, AI might also bring us peril, as it poses an existential threat to our continued existence. What if we create a greater intelligence than our own, and that intelligence dominates the Earth, perhaps wiping out all humanity?

While existential threats may not trouble most of us greatly, they trouble transhumanists enormously. Such threats stand between a transhumanist and his immortality. I say '*his* immortality' as transhumanists seem to be predominately male. Nick Bostrom, for example, has argued of the immense loss of happiness to all those not yet born, who might never live if humanity is eliminated. However, the end of humanity would also be a significant personal loss for someone like Bostrom, who desires to live forever.

Elon Musk, in discussing his plans to colonise Mars, has quipped that his ambition is to die on Mars, but not on impact. But I suspect the truth of the matter is that he doesn't want to die at all – on Earth, on Mars or indeed among the stars. Musk's reasons for going to Mars rest on his belief that life on Earth is doomed in the long term. The Earth is heading for ecological collapse. And it is technologically too soon to go to the stars. Mars offers the only lifeboat in our solar system that might save humanity.

Musk is investing a large chunk of his wealth in advancing artificial intelligence. In 2015, he, Sam Altman and other founders pledged to give over $1 billion to the OpenAI organisation. In 2019, Microsoft invested a further $1 billion in OpenAI. The original goals of OpenAI were to build an 'artificial general intelligence' – an AI that surpasses humans in all aspects of their intelligence – and to share it with the world. However, the part about sharing it with the world now seems rather doubtful. In 2019, OpenAI transformed itself from a not-for-profit into a limited partnership with an exotic capped-profit limit of 100 times an investment.

Musk has also invested $100 million of his wealth in the Neuralink Corporation. This is a neuro-technology company developing a 'neural lace', an implantable brain–machine interface.

Musk has argued that the only way we'll be able to keep up with the machines as they become ever more intelligent is by connecting our brains directly to them. The logic of this argument seems somewhat dubious to me. I fear it may reveal to those machines just how limited we are ...

Transhumanism thus offers a somewhat disturbing vision of our future, a future where we extend ourselves with machines, and perhaps even upload ourselves onto them. And this is, at least in part, the vision driving a number of people working to advance AI.

WISHFUL THOUGHTS

Besides wanting to live forever, there are other, more down-to-earth traits that trouble the people working in Silicon Valley. One is a techno-optimistic form of confirmation bias, the well-known problem in human decision-making by which people will seek out and remember information that confirms their beliefs, but overlook and forget information that disagrees with them.

Perhaps it's not surprising that a place like Silicon Valley is full of people prone to confirmation bias, especially when it comes to the power of technology to shape a positive future. Wishful thinking is likely to emerge in a town awash with money, where millions and even billions of dollars are to be made in remarkably little time from what are sometimes surprisingly simple technological innovations. For those who have been fortunate enough to win those millions and billions, I imagine it must be hard to avoid.

It's easy when you've got so lucky to believe wrongly in the superiority of your decision-making, and in the power of digital technologies, as opposed to the luck of the draw. And with more lucky millionaires and billionaires per square kilometre than anywhere else on the planet, the Valley has a particular problem of techno-optimistic confirmation bias.

The sorry tale of the biomedical company Theranos and its Steve Jobs–inspired founder, Elizabeth Holmes, is a good example of how people in the Valley hear what they want to hear and disregard everything else. There was plenty of evidence from early on in the company's 15-year history that things were not quite right.

'Edison', the revolutionary blood-testing device that the company had invented, and which needed just a few drops of blood to scan for disease, catapulted Holmes onto the front pages of *Forbes* and *Fortune* magazines. It was an invention that we all wanted to exist. Who, after all, likes to have blood taken? And the company raised more than $700 million from venture funds and private investors on the back of this dream.

Theranos claimed it needed just 1/1000th of the amount of blood that would normally be needed for a blood test. In reality, the only 1/1000th that should have been claimed was that the company's annual income was 1/1000th of the $100 million written in press releases. 'Edison' never actually worked. Indeed, it's not at all clear it could ever have worked as claimed without violating the laws of physics.

Despite this, Theranos quickly became one of the Valley's most talked about 'unicorns', meaning a privately held company valued at over $1 billion. At its peak in 2015, Theranos was valued at around $10 billion. Elizabeth Holmes was briefly the youngest self-made female billionaire in the world, and the wealthiest in the United States. Today, Holmes is awaiting sentencing after being found guilty of four counts of fraud. If convicted, she might spend 20 years in jail. Theranos is arguably the largest biomedical fraud in history. How could it be that this house of cards didn't come crashing down sooner?

The board of Theranos included many politically savvy people, such as George Schultz, a former director of the United States Office of Management and Budget; James N. Mattis, a former Secretary of Defense; Senator Sam Nunn, a former chair of the powerful Senate Armed Services Committee; and Henry Kissinger, a former Secretary of State. You wouldn't expect experienced politicians like this to be easily deceived. But they were.

Investors in Theranos included Tim Draper, of the legendary venture capital firm Draper Fisher Jurvetson, the Walton family, and media baron Rupert Murdoch. Draper Fisher Jurvetson has invested in many success stories, like Tesla, SpaceX and Twitter, while the Waltons are America's richest family. You wouldn't expect experienced investors like this to back a fraud. But they did.

Perhaps most surprisingly, the medical advisory board of Theranos included past presidents or board members of the American Association for Clinical Chemistry, and a former director of the Centers for Disease Control and Prevention, which is the leading US public health institution. The board members were invited to review the company's proprietary technologies and advise on their integration into clinical practice. But they failed to raise a red flag in time to avoid this scandal.

In somewhere like the Valley, it's easy to believe the dream and ignore the naysayers. Technology will solve whatever problems challenge you. And technology will make lots of people millions – and in some cases billions – of dollars in the process. This, then, is part of the rather dangerous mindset of many people building AI.

THE TENDERLOIN

In downtown San Francisco, young coders building this AI future walk past homeless people sleeping rough on the streets. The contrast is perhaps most stark in the Tenderloin, the famously gritty neighbourhood of downtown San Francisco, close to Union Square.

San Francisco has around 7000 homeless people. This is the third-highest rate per capita of any city in the United States. The UN special rapporteur Leilani Farha has compared it to Mumbai. There are similar numbers of homeless people further south in Silicon Valley. The waitlist to get into a homeless shelter in San Francisco on any night contains over 1200 people.

This housing crisis is amplified by the citywide boom in real estate, which is driven in part by the inflated salaries paid by technology companies, as well as by the stratospheric initial public offerings (or IPOs) of many of these companies. The average house in San Francisco now costs over $1.4 million, the most expensive of any city in the United States. Houses in San Francisco spend just 16 days on the market before being sold. House prices have more than tripled in the last 20 years.

Average wages, on the other hand, have not tripled. Indeed, they have barely kept up with inflation. Not surprisingly, buying a house is therefore out of reach for many in San Francisco. Renting isn't much better. The median rent for a one-bedroom apartment in San Francisco is over $3700 a month, the highest of any US city. This median rent is nearly $1000 more than in New York City.

So how do you fix San Francisco's terrible homelessness problem? Code Tenderloin, a non-profit based in San Francisco, believes it is by teaching homeless people to code. They offer a six-week/72-hour course which teaches the basics of JavaScript using software developers from local tech companies as course instructors.

Many countries around the world, on the other hand, have discovered a much simpler and more direct way to solve homelessness: build more affordable housing. Yes, it's that simple – just make housing more accessible. Finland, as an example, is the only country in Europe where homelessness is falling. It gives homeless people homes *unconditionally*, as soon as they need them. And the state then supports them as they deal with the mental health, drug and other issues common to those who live on the streets.

Giving people homes is a simple and effective solution to homelessness. It's hard not to contrast this with the technophiles, who think the solution to homelessness is teaching people Javascript.

PROJECT MAVEN

At this point, I imagine you are justifiably concerned about the sort of people building this AI future. Despite the concerns I've raised, not all is doom and gloom. And there are some green shoots that suggest optimism. Many people working in AI are waking up to their significant ethical responsibilities. Indeed, this book is a small part of my own awakening.

Perhaps one of the clearest examples of the emerging moral conscience in the AI research community is the pushback that has occurred around Google's involvement in Project Maven. This is a controversial Pentagon project that, starting in April 2017, developed machine-learning algorithms to identify people and objects in video images. Military intelligence analysts are drowning in a sea of imagery collected by drones and satellites. Could AI help tame this torrent of data?

There is little wrong with having AI support human analysts as they interpret such imagery. In fact, there are ethical arguments that computer-vision algorithms might help prevent mistakes like the May 1999 bombing of the Chinese Embassy in Belgrade, which killed three Chinese people and wounded 20 others. Satellite-guided bombs from a B-2 stealth bomber struck the embassy during the Kosovo War because CIA analysts had misidentified the building. Perhaps a second pair of computer eyes might have helped avoid this error?

The problem with this idea is that the same computer-vision algorithms can be used to entirely automate the identification and selection of targets by drone, turning what today is a semi-autonomous platform into a fully autonomous weapon. Indeed, in early 2020 Turkey started deploying drones on its border with Syria which are believed to use facial-recognition software to identify, track and kill human targets on the ground, all without human intervention. There is much to be concerned about ethically with such drones.

Many of Google's employees were understandably worried by the company's foray into military research, especially given the secrecy behind the project. Google's motto for many years was 'Don't be evil'. Building software that could eventually lead to fully autonomous drones perhaps challenges that idea. Over 4000 Google employees, including dozens of senior engineers, signed a petition in protest against the company's involvement. At least a dozen employees quit. Google responded by announcing it would not be continuing with Project Maven. And in 2019 Google did indeed choose not to renew its contract with the Department of Defense.

This controversy encouraged Google to take more seriously the growing public concerns around its development and deployment of AI. In June 2018, it released its principles for the responsible use of AI. Nine months later, in March 2019, Google announced an AI and ethics board. However, conflict over its membership led to the board being shut down a week later. In August 2020, Google announced that it would start selling ethics advice to other companies looking to use AI.

As for Project Maven: the $111-million contract with the Department of Defense was picked up by Peter Thiel's controversial AI software developer Palantir Technologies. Which elegantly brings me to the next chapter, in which I discuss not the people building AI, but the companies.

THE COMPANIES

THE NEW TITANS

The AI revolution is being led by companies that don't follow, and often don't answer to, the usual corporate rules. This has created some fundamental problems, often of an ethical nature, and it will continue to do so.

Many previous scientific revolutions started in universities and government laboratories. Penicillium was discovered, for example, by Sir Alexander Fleming in 1928 at St Mary's Hospital Medical School, then part of the University of London. It was a serendipitous discovery, and the medical revolution that followed undoubtably changed the planet.

As a second example, the double-helix structure of DNA was discovered by James Watson and Francis Crick in 1953 at the University of Cambridge, using data collected by Maurice Wilkins and Rosalind Franklin at King's College London. This was an important step in unlocking the mysteries of life. The genetic revolution that followed has only just begun to change our lives.

Perhaps most relevant to our story, the first general-purpose digital computer was built in 1945 at the University of Pennsylvania's

Moore School of Electrical Engineering. It was a 30-ton hulk made up of 18,000 vacuum tubes, over 7000 crystal diodes, 1500 relays and 10,000 capacitors that consumed 150 kilowatts of electric power. But it was a thousand times faster than the electro-mechanical calculators it replaced. The computing revolution that followed has most definitely changed our lives greatly.

While artificial intelligence also started out in universities – at places like MIT, Stanford and Edinburgh during the 1960s – it is technology companies such as Google, Facebook, Amazon and IBM, as well as younger upstarts like Palantir, OpenAI and Vicarious, that are driving much of the AI revolution today. These companies have the computing power, data sets and engineering teams that underpin many of the breakthroughs, and that numerous academic researchers like me unashamedly envy.

The largest of these companies are rightly known as 'Big Tech'. But not because they employ lots of people. Indeed, for every million dollars of turnover, they employ roughly 100 times fewer people than companies in other sectors. Facebook, for example, has over 120 times fewer employees per million dollars of turnover than McDonald's.

The impressive market capitalisation of the Big Tech companies is one reason for the moniker 'big'. Their share prices are truly spectacular, a concentration of wealth that the world has never seen before. The first trillion-dollar company in the history of the planet was Apple. It crossed over to a market cap of 13 figures in August 2018. Two years later, Apple had doubled in value. Apple is now worth more than the whole of the FTSE 100 Index, the

100 most valuable companies listed on the stock market in the United Kingdom.

Since Apple became a trillion-dollar company, three other technology companies have joined the four-comma club: Amazon, Microsoft and Alphabet (the parent company of Google). Facebook is likely to join these trillion-dollar stocks shortly. The immense wealth of these companies gives them immense power and influence. Governments around the world are struggling to contain them. In 2019, Amazon had a turnover of over $280 billion. That is more that the GDP of many small countries. For example, Apple's turnover is more than the GDP of Portugal ($231 billion in 2020), and nearly 50 per cent more than that of Greece ($189 billion). Amazon's turnover puts the productivity of its 1 million employees on a par with the 5 million people of Finland, who together generated $271 billion of wealth in 2020.

The Big Tech companies dominate their markets. Google answers eight out of every nine search queries worldwide. If it weren't effectively locked out of China, it would probably answer even more. The other Big Tech companies are also dominant in their own spaces. Two billion out of the nearly 8 billion people on the planet use Facebook. In the United States, Amazon is responsible for around half of all e-commerce. And in China, Alibaba's payment platform Alipay is used for about half of all online transactions.

The founders of technology companies, large and small, are unsurprisingly celebrated like rock stars. We know many of them by their first names. Bill and Paul. Larry and Sergey. Mark and

Jeff. But they are modern-day robber barons just like Mellon, Carnegie and Rockefeller, the leaders of the technological revolution of their time.

Many of these founders wield huge power. This goes well beyond the power that CEOs of companies in other sectors typically possess. In part, this is to be welcomed. It has enabled innovation and allowed technology companies to move fast. But in moving fast, many things have been broken.

One reason for such power is the unconventional share structures of technology companies. Even when they have given up majority ownership of their companies, many have retained absolute or near-absolute decision-making power. They can easily push back against any resistance.

As an example, Facebook's Class B shares have ten times the voting rights of the Class A shares. And Mark Zuckerberg owns 75 per cent of Facebook's Class B shares. He has therefore been able to ignore calls from other shareholders for Facebook to reform. And for reasons that are hard to understand, the Securities and Exchange Commission has no problem with founders like Zuckerberg being both Facebook's CEO and the chair of its board.

Perhaps most egregious was the listing on the New York Stock Exchange of Snap Inc., the company behind Snapchat, in March 2017. This IPO sold shares to the public that had no voting rights at all. Despite this, the IPO raised $500 million. And the stock closed its first day up 44 per cent. What was the Securities and Exchange Commission thinking? How did we go from executives of publicly listed companies being accountable to the shareholders, to executives being unaccountable to anyone but themselves? And how did investors learn to not care?

NOTHING VENTURED

Another reason technology companies are so powerful is that the market doesn't expect them to be profitable. Unicorns like Uber, Snapchat and Spotify have, for example, never turned a profit. Even those that do are not expected to return much to investors. This is ironic, as they can better afford to return dividends to their shareholders than many companies that traditionally do.

Most Big Tech companies are sitting on large cash mountains. It is estimated that US companies have over $1 trillion of profits waiting in offshore accounts for a tax break to bring them home. Apple is one of the worst offenders, with around $250 billion sitting offshore. But at least it pays out a small cash dividend, giving shareholders a yield of around 1 per cent. Amazon is sitting on around $86 billion in cash and has annual profits of around $33 billion, yet has never returned any of this to its investors in the form of a dividend.

Even those technology companies that haven't been able to return a profit have had little difficulty in raising billions of dollars from investors. Uber, for example, got over $20 billion when it went public in 2019. In that same year, it lost $1 out of every $4 it earned in its revenue of $14 billion. Indeed, it's not clear to me if Uber can ever be profitable.

The typical view within technology companies is that it is better to invest for growth and gain market share, rather than being a profit-making business in the short term. There is some truth to this growth-at-all-costs strategy – Amazon has demonstrated its long-term value. But for every Amazon, there are poorly managed companies like Pets.com, the poster child of the dotcom crash, that were never going to be sustainable.

The problem begins with the ease with which technology companies can raise money. Venture capital distorts markets. Why haven't drivers banded together and formed a cooperative to compete against Uber? Ideally, we should use technology to connect drivers seamlessly with passengers, to create a friction-less marketplace. That is the brilliance of Uber's business model. But at the end of the day, Uber is largely a tax sitting in between those wanting a ride and those who can offer a ride. Uber is stealing much of the value out of the system. Uber drivers who earn so little they have to sleep in their cars are certainly not getting sufficient value out of the market. Digital technologies are meant to get rid of friction. In the long term, the ultimate winner in the ride-sharing market is neither the passenger nor the driver. The winner is simply the venture fund with the deepest pockets.

A driver cooperative can't compete against a business like Uber that doesn't need to make money. And certainly not against a business like Uber that doesn't even need to break even, but will happily lose money year after year until its competitors have been driven out of business.

Such factors have made technology companies an alarming force in the development and deployment of artificial intelligence. Awash with cheap money. Lacking in transparency and accountability. Engineered to disrupt markets. And driven by idiosyncratic founders with near absolute power. It's hard to think of a more dangerous cocktail.

SUPER-INTELLIGENCE

One somewhat distant concern about artificial intelligence is the threat posed by the emergence of super-intelligence. From what I can tell, most of my colleagues, other researchers working in AI, are not greatly worried about the idea that we might one day build super-intelligent machines. But this possibility has tortured many people outside the field – like the philosopher Nick Bostrom.[1]

One of Bostrom's fears is that super-intelligence poses an existential threat to humanity's continuing existence. For example, what if we build a super-intelligent machine and ask it to make paperclips? Might it not use its 'superior' intelligence to take over the planet and turn everything, including us, into paperclips?

This is what is called a 'value alignment problem'. The values of this super-intelligent paperclip-making machine are not properly aligned with those of humankind. It's very difficult to specify precisely what we would like a super-intelligence to do. Suppose we want to eliminate cancer. 'Easy,' a super-intelligence might decide: 'I simply need to get rid of all hosts of cancer.' And so it would set about killing every living thing!

One reason I don't have existential fears about some non-human super-intelligence is that we already have non-human super-intelligence on Earth. We already have a machine more intelligent than any one of us. A machine with more power and resources at its disposal than any individual. It's called a company.

Companies marshal the collective intelligence of their employees to do things that individuals alone cannot do. No individual on their own can design and build a modern microprocessor. But Intel can. No individual on their own can design and build a nuclear power station. But General Electric can.

Probably no individual on their own will build an artificial general intelligence, an intelligent machine that matches or even exceeds any human intelligence. But it is highly likely that a company will, at some point in the future, be able to do so. Indeed, as I say, companies already are a form of super-intelligence.

That brings me neatly back to the problem of value alignment. This seems precisely to be one of the major problems we face today with these super-intelligent companies. Their parts – the employees, the board, the shareholders – may be intelligent, ethical and responsible. But the behaviours that emerge out of their combined super-intelligent efforts may not be ethical and responsible. So how do we ensure that corporate values are aligned with the public good?

THE CLIMATE EMERGENCY

If you will indulge me, I'm going to make a little detour to the topic of the climate emergency. Here we have perhaps the clearest example of an issue where the values of companies have not been aligned with the public good. On a positive note, however, there is some evidence that in recent years those corporate values are starting to align with the public good.

More than a century ago, in 1899, the Swedish meteorologist Nils Ekholm suggested that the burning of coal could eventually double the concentration of atmospheric CO_2, and that this would 'undoubtedly cause a very obvious rise of the mean temperature of the Earth'. Back then, the fear was mostly of a new ice age. Such an increase in temperature was therefore considered desirable: a way of preventing this eventuality.

However, by the 1970s and 1980s, climate change had become a subject of serious scientific concern. These concerns culminated in the setting up of the Intergovernmental Panel on Climate Change (IPCC) by the United Nations in 1988. The IPCC was established to provide a scientific position on climate change, as well as on its political and economic impacts.

Meanwhile, oil companies such as Exxon and Shell were also researching climate change. In July 1977, one of Exxon's senior scientists, James Black, reported to the company's executives that there was general scientific agreement that the burning of fossil fuels was the most likely cause of global climate change. Five years later, the manager of Exxon's Environmental Affairs Program, M.B. Glaser, sent an internal report to management that estimated that fossil fuels and deforestation would double the carbon dioxide concentrations in the Earth's atmosphere by 2090. In fact, CO_2 concentrations in the Earth's atmosphere have already increased by around 50 per cent. The best estimates now are that a doubling will occur even sooner, perhaps by 2060.

Glaser's report suggested that a 'doubling of the current [CO_2] concentration could increase average global temperature by about

1.3 degrees Celsius to 3.1 degrees Celsius', that 'there could be considerable adverse impact including the flooding of some coastal land masses as a result of a rise in sea level due to melting of the Antarctic ice sheet', and that 'mitigation of the "greenhouse effect" will require major reductions in fossil fuel combustion'.

Despite these warnings, Exxon invested significant resources in climate change *denial* during the decades that followed. Exxon was, for example, a founding member of the Global Climate Coalition, a collection of businesses opposed to the regulation of emissions of greenhouse gases. And Exxon gave over $20 million to organisations denying climate change.

It wasn't until 2007 that ExxonMobil (as the company had become) publicly acknowledged the risks of climate change. The company's vice president for public affairs, Kenneth Cohen, told the *Wall Street Journal* that 'we know enough now – or, society knows enough now – that the risk is serious and action should be taken'. However, it would take seven more years before ExxonMobil released a report acknowledging the risks of climate change for the first time.

Clearly, the values of ExxonMobil were not aligned with those of the wider public. But there is some evidence that corporate values are starting to shift in a favourable direction. ExxonMobil has, for instance, invested around $100 million in green technologies. The company now supports a revenue-neutral tax on carbon, and has lobbied for the United States to remain in the Paris Climate Agreement.

Many other companies are starting to act on climate change. The news was rather overshadowed by the unfolding pandemic, but in February 2020 the oil and gas giant BP promised to become carbon-neutral by 2050 or sooner. Also in February 2020, Australia's second-largest metals and mining company, Rio Tinto, announced it would spend $1 billion to reach net zero emissions by 2050.

In July 2020, Australia's largest energy provider, and the country's biggest carbon emitter, AGL Energy, announced a target of net zero emissions by 2050. It even tied the long-term bonuses of its executives to that goal. And in September 2020, the world's largest mining company, BHP, joined the club, planning to be net zero by 2050. While politicians are in general failing to act with sufficient urgency on the climate emergency, it is good to see that many companies are finally prepared to lead the way.

Indeed, it is essential that companies act. Since the founding of the IPCC in 1988, just 100 companies have been responsible for 71 per cent of greenhouse gas emissions globally. ExxonMobil is the fifth-most polluting out of these 100 corporations: it alone produces 2 per cent of all global emissions. We can make personal changes to reduce our carbon footprint, but none of that will matter if these 100 companies don't act more responsibly.

BAD BEHAVIOUR

Let's return to the tech sector. There is plentiful evidence that technology companies, just like the 100 companies responsible for the majority of greenhouse gas emissions, have a value alignment problem. You could write a whole book about the failures of technology companies to be good corporate citizens. I'll just give you a few examples, but new ones are uncovered almost every day.

Let's begin with Facebook's newsfeed algorithm. This is an example of a value alignment problem on many levels. On the software level, its algorithm is clearly misaligned with the public good. All Facebook wants to do is maximise user engagement. Of course, user engagement is hard to measure, so Facebook has decided instead to maximise clicks. This has caused many issues. Filter bubbles. Fake news. Clickbait. Political extremism. Even genocide.[2]

Facebook's newsfeed algorithm is also an example of a value alignment problem at the corporate level. How could it be that Facebook decided that clicks were the overall goal? In September 2020, Tim Kendall, who was 'Director of Monetization' for Facebook from 2006 until 2010, told a Congressional committee:

> We sought to mine as much attention as humanly possible ...
> We took a page from Big Tobacco's playbook, working to make
> our offering addictive at the outset ... We initially used engage-
> ment as sort of a proxy for user benefit. But we also started to
> realize that engagement could also mean [users] were sufficiently
> sucked in that they couldn't work in their own best long-term
> interest to get off the platform ... We started to see real-life con-
> sequences, but they weren't given much weight. Engagement
> always won, it always trumped.[3]

In 2018, as evidence of the harmful effects of Facebook's newsfeed algorithm became impossible to ignore, Mark Zuckerberg announced a major overhaul: the newsfeed would now emphasise 'meaningful social interactions' over 'relevant content'. The changes prioritised content produced by a user's friends and family over 'public content', such as videos, photos or posts shared by businesses and media outlets.

Facebook's corporate values are arguably in opposition to the public good in a number of other areas too. In October 2016, for example, the investigative news outlet *ProPublica* published a story under the headline 'Facebook Lets Advertisers Exclude Users by Race'.[4] The story exposed how Facebook's micro-targeting tools let advertisers direct adverts at its users[5] according to their race and other categories.

Adverts for housing or employment that discriminate against people based on race, gender or other protected features are prohibited by US federal law. The *Fair Housing Act* of 1968 bans adverts that discriminate 'based on race, color, religion, sex, handicap, familial status, or national origin'. And the *Civil Rights Act* of 1964 prohibits job adverts which discriminate 'based on race, color, religion, sex and national origin'.

Despite the outcry that followed *ProPublica*'s story, Facebook continued to let advertisers target their adverts by race. One year later, in November 2017, *ProPublica* ran the headline 'Facebook (Still) Letting Housing Advertisers Exclude Users by Race'.[6] Nothing much had changed. As a computer programmer myself, I can't believe that it takes years to remove some functionality from the part of Facebook's code that sells adverts. Facebook has 45,000 employees to throw at the problem. I can only conclude that the company doesn't care. And that the regulator didn't make it care.

I could pick on many other technology companies that have demonstrated values misaligned with the public good. Take Google's YouTube, for instance. In 2019, Google was fined $170 million by the US Federal Trade Commission (FTC) and New York's attorney-general for violating children's privacy on YouTube. The *Children's Online Privacy Protection Act (COPPA)* of 1998 protects children under the age of 13, meaning parental consent is required before a company can collect any information about a child.

Google knowingly violated *COPPA* by collecting information about young viewers of YouTube. There are over 5 million subscribers of its 'Kids Channel', most of whom, it seems fair to guess, are children. And many of the 18.9 million subscribers to its *Peppa Pig* channel are also probably children. But Google collects information about these subscribers to engage them longer on YouTube, and to sell adverts.

Google boasted to toy companies such as Mattel and Hasbro that 'YouTube was unanimously voted as the favorite website for kids 2-12', and that '93% of tweens visit YouTube to watch videos'. Google even told some advertisers that they did not have to comply with *COPPA* because 'YouTube did not have viewers under 13'. YouTube's terms of service do indeed require you to be over 12 years old to use the service. But anyone with kids knows what a lie it is for Google to claim that YouTube does not have child viewers.

The $170-million fine was the largest fine the FTC has so far levelled against Google. It is, however, only a fraction of the $5-billion fine the FTC imposed on Facebook earlier in 2019, in response to the privacy violations around Cambridge Analytica. This matched the $5-billion fine that the EU imposed on Google for antitrust violations connected to its Android software.

The case the FTC brought against Google is not the end of the YouTube matter. A new lawsuit was filed in a UK court in September 2020, claiming that YouTube knowingly violated the United Kingdom's child privacy laws; it is seeking damages of over $3 billion. You have to wonder how big the fines need to be for Big Tech to care.

CORPORATE VALUES

Technology companies often have rather 'wacky' values. Some of this is marketing. But it also tells us something about their goals, and how they intend to pursue those goals.

Paul Buchheit – lead developer of Gmail and Google employee number 23 – came up with Google's early motto, 'Don't be evil'. In a 2007 interview, he suggested, with not a hint of irony, that it's 'a bit of a jab at a lot of the other companies, especially our competitors, who at the time, in our opinion, were kind of exploiting the users to some extent'. He also claimed that he 'wanted something that, once you put it in there, would be hard to take out'.[7]

At least this part of his prediction has come true: 'Don't be evil' is still in Google's code of conduct. But only just. In 2018, 'Don't be evil' was relegated from the start to the final sentence on the 17th and last page of Google's rather long and rambling code of conduct.

Facebook's motto was perhaps even more troubling: 'Move fast and break things'. Fortunately, Facebook realised that breaking too much in society could be problematic, and so in 2014 the motto was changed to 'Move fast with stable infrastructure'. Not so catchy, but it does suggest a little more responsibility. This is to be welcomed as, for most of its corporate life, moving fast and breaking things seemed an excellent summary of Facebook's troubled operations.

Amazon's mission slogan is also revealing: 'We strive to offer our customers the lowest possible prices, the best available selection, and the utmost convenience. To be Earth's most customer-centric company, where customers can find and discover anything they might want to buy online.' Amazon is clearly laser-focused on its customers. So it's not surprising, then, that news stories regularly reveal how workers in its distribution centres are being poorly treated. Or how suppliers are being squeezed by the tech giant.

Microsoft's mission is 'to empower every person and organization on the planet to achieve more'. But GoDaddy, the domain name register and web-hosting site, is even more direct: 'We are here to help our customers kick ass.' Clearly, value alignment in Big Tech is a problem we need to address.

GOOGLE'S PRINCIPLES

Following the controversy around Project Maven, Google drew up seven principles that would guide its use of AI.[8] These principles were made public in June 2018.

We believe that AI should:

1. Be socially beneficial.

2. Avoid creating or reinforcing unfair bias.

3. Be built and tested for safety.

4. Be accountable to people.

5. Incorporate privacy design principles.

6. Uphold high standards of scientific excellence.

7. Be made available for uses that accord with these principles.

In addition to the above principles, Google also promised not to 'design or deploy AI in the following application areas':

1. Technologies that cause or are likely to cause overall harm ...

2. Weapons or other technologies whose principal purpose or implementation is to cause or directly facilitate injury to people.

3. Technologies that gather or use information for surveillance violating internationally accepted norms.

4. Technologies whose purpose contravenes widely accepted principles of international law and human rights.

The last promise seems entirely unnecessary. Contravening international law is already prohibited. By law. Indeed, the fact that many of these ethical principles needed to be spelled out is worrying. Creating unfair bias. Building technology that is unsafe, or lacking in accountability. Causing harm to your customers. Damaging human rights. No business should be doing any of this in the first place.

Even if Google's AI principles sound good on paper, a couple of significant problems remain. In particular, who is going to police Google? And how can we be sure they will? Nine months after the

AI principles were first published, in March 2019, Google established the Advanced Technology External Advisory Council (ATEAC) to consider the complex challenges arising out of its AI principles. The council had eight members. It was slated to meet four times each year to consider ethical concerns around issues like facial recognition, the fairness of machine-learning algorithms, and the use of AI in military applications.

There was, however, an immediate outcry about this AI ethics board, both from within and outside Google. One of the eight council members was Kay Coles James, president of the Heritage Foundation, a conservative think-tank with close ties to the administration of then president Donald Trump. Thousands of Google employees signed a petition calling for her removal over what they described as 'anti-trans, anti-LGBTQ and anti-immigrant' comments she had made.

One board member, Alessandro Acquisti, a professor of Information Technology and Public Policy at Carnegie Mellon University, quickly resigned, tweeting: 'While I'm devoted to research grappling with key ethical issues of fairness, rights and inclusion in AI, I don't believe this is the right forum for me to engage in this important work.'

Another board member tweeted: 'Believe it or not, I know worse about one of the other people.' This was hardly a glowing endorsement of Kay Coles James. Indeed, it was more of an indictment of another (unnamed) member of the board.

One week after announcing the external ethics board, and after a second board member had resigned, Google accepted the inevitable. It closed the board down and promised to think again about how it policed its AI principles. At the time of writing, in February 2022, Google has yet to announce a replacement.

IBM'S THINKING

Around the same time that Google was grappling with its AI principles, IBM offered up its own guidelines for the responsible use of AI. It proposed these principles both for itself and as a roadmap by which the rest of industry could address concerns around AI and ethics.

IBM's Principles for Trust and Transparency

1. The purpose of AI is to augment human intelligence.

2. Data and insights belong to their creator.

3. New technology, including AI systems, must be transparent and explainable.[9]

Unfortunately, all three principles seem poorly thought out. It's as though IBM had a meeting to decide on some AI principles, but the meeting wasn't scheduled for long enough to do a proper job. Let me go through each principle in turn.

Yes, AI can 'augment' human intelligence. AI can help us be better chess players, radiologists or composers. But augmenting human intelligence has never been AI's sole purpose. There are many places where we actually want artificial intelligence to *replace* human intelligence entirely. For example, we don't want humans doing a dangerous job like clearing a minefield, when we can get robots to do it instead. Less dramatically, we don't want human intelligence doing a dull job like picking items in a warehouse, when we can get robot intelligence to do it for us.

There are also places where we want artificial intelligence not to augment human intelligence but to *conflict* with human intelligence. The best way to surpass human intelligence might not be to try to extend what we can do, but to go about things in completely new ways. Aeroplanes don't augment the natural flight of birds. They go about heavier-than-air motion in an entirely different way.

We've built many tools and technologies that don't 'augment' us. X-ray machines don't augment our human senses. They provide an entirely new way for us to view the world. Similarly, AI won't always simply extend human intelligence. In many settings, it will take us to totally new places. It is misguided, then, to say that AI is only ever going to 'augment' us, or that AI is never going to take over some human activities completely. At best, it is naive; at worst, it is disingenuous. It comes across to me as a clumsy attempt to distract attention from fears about jobs being replaced.

Asserting that data and insights belong to the creators – IBM's second ethical principle – may seem like a breath of fresh air in an industry notorious for stealing data from people and violating their privacy. However, international laws surrounding intellectual property already govern who owns such things. We don't need any new AI principles to assert these rights. Again, this seems a somewhat clumsy attempt to distract from the industry's failure to respect such data rights previously. Indeed, claiming that 'data and insights belong to their creator' is itself an ethical and legal mine-field. What happens when that creator *is* an AI? IBM, do you really want to open that Pandora's box?

Finally, IBM's third ethical principle is that AI systems must be transparent and explainable. While it's often a good idea for AI systems to be transparent and explainable, there are many settings where this simply might not be possible. Let's not forget that the alternative – human decision-making – is often not very transparent. Transparency can even be a bad thing. It may enable bad people to hack you.

As for explanation: we cannot get a computer-vision system today to explain how its algorithms recognise a stop sign. But then, even the best neurobiologist struggles to explain how the human eye and brain actually see a stop sign. It's entirely possible that we'll never be able to explain adequately either human or artificial sight.

Transparency and explainability help build trust, but, as I shall argue later, there are many other components of trust, such as fairness and robustness. Transparency and explainability are a means to an end. In this case, the end is trust. But – and this is where IBM gets it wrong – transparency and explainability are not an end in themselves.

What good would there be in an AI system that transparently explained that it was not offering you the job because you were a woman? Unless you have money and lawyers, there would likely be nothing you could do about this harm. A much better third principle would have been that any new technology, AI included, should be designed and engineered to be *worthy* of your trust.

RETHINKING THE CORPORATION

Given all these concerns about technology companies, and their faltering steps to develop and deploy AI responsibly, it is worth considering how we might change things for the better. It is easy to forget that the corporation is a relatively new invention, and very much a product of the Industrial Revolution. Most publicly listed companies came into being only very recently. And many will be overtaken by technological change in the near future.

Fifty years ago, companies on the S&P 500 Index lasted around 60 years. Today, most companies on the index last only around two decades. It is predicted that three-quarters of the companies on the S&P 500 today will have disappeared in ten years' time.

It is also worth remembering that the corporation is an entirely human-made institution. It is designed in large part to permit society to profit from technological change. Corporations provide the scale and coordination to build new technologies. Limited liability lets directors take risks with new technologies and new markets without incurring personal debt. And venture capital, along with bond and equity markets, gives companies access to funds that enable them to invest in new technologies and expand into new markets.

Twenty years ago, only two of the top five companies in the world were tech companies. The industrial heavyweight General Electric was the most valuable publicly listed company, followed by Cisco Systems, ExxonMobil, Pfizer and Microsoft. Today, all five of the most valuable listed companies are digital technology companies: Apple, Microsoft, Amazon, Alphabet and Facebook. Close behind them are Tencent and Alibaba.

Perhaps it is time to think, then, about how we might reinvent the idea of the corporation to better suit the ongoing digital revolution. How can we ensure that corporations are more aligned with the public good? How do we better share the spoils of innovation?

There was another invention of the Industrial Revolution that was designed to meet many of these ends. But, sadly, it seems to be dying away. This was the mutual society. Unfortunately, mutuals have some competitive disadvantages. For example, until recently, mutuals were unable to raise capital other than by retaining past profits. This has put them at a severe disadvantage to publicly listed companies.

A newer invention may be part of the answer: the B (or benefit) corporation. This is a purpose-driven for-profit business that creates benefit for all stakeholders, not just shareholders. A business that balances purpose and profit by considering the interests of workers, customers, suppliers, community and the environment, and not just those of shareholders.

There are over 3300 certified B corporations today across 150 industries in over 70 countries. Household names like Ben & Jerry's (ice cream makers) and Patagonia (outdoor apparel) are certified B corporations. But as yet there is only one B corporation I'm aware of that is focused on AI.

Lemonade is a for-profit B corporation that is using artificial intelligence to disrupt the staid insurance industry. It offers homeowners, renters and pet owners insurance in the United States. It returns underwriting profits to non-profits selected by its community during its annual giveback. The company aims to do well in its business by doing good in its community.

Lemonade uses AI chatbots and machine learning to super-charge and automate much of the customer experience. It takes just 90 seconds to get insured, and a mere three minutes to pay many claims. If you believe the marketing, the company is much loved by its customers. Perhaps we need a lot more such B corporations developing AI responsibly and giving back to their communities?[10]

There was one other high-profile non-profit set up to develop AI responsibly. OpenAI was founded in San Francisco in late 2015 by Elon Musk and some other investors. They collectively pledged to invest a cool $1 billion in OpenAI to ensure that AI benefits all of humanity. In 2019, Microsoft invested an additional $1 billion. But at the same time, OpenAI stopped being a non-profit company.

In May 2020, OpenAI announced the world's largest neural network, a 175-billion-parameter language neural network called GPT-3. Many AI experts were blown away by the neural network's ability to generate stories, write poems, even generate simple computer code. Three months later, in September 2020, OpenAI announced it had *exclusively* licensed GPT-3 to Microsoft.

I'm still waiting for OpenAI to announce that its name has changed to ClosedAI. It's hard to see OpenAI delivering on its goal of ensuring that AI benefits all of humanity. In the meantime, we'll continue to worry about the ethics of the companies, as well as the people, building AI.

So far I have focused on the people and companies building AI, as their goals and desires are reflected in some of the AI that is being created today. I turn next to some of the pressing ethical issues surrounding AI's development and deployment.

AUTONOMY

A NEW CHALLENGE

I begin with what is perhaps the *only* property that makes AI special. This is autonomy. We're giving machines the ability to act in our world somewhat independently. And these actions can have an impact on humans. Not surprisingly, autonomy raises a whole host of interesting ethical challenges.

New technologies often introduce ethical challenges. And you could argue that autonomy is the one new ethical challenge that AI poses. All the other challenges that AI introduces, such as bias or invasion of our privacy, are ones we have faced before. For example, we have been trying to tackle racial bias for decades. AI may have put the problem on steroids, but it is an old problem. Similarly, many governments have been encroaching on our personal privacy for decades. AI may have put this problem on steroids too, but it is not a new problem.

Autonomy, on the other hand, is an entirely novel problem. We've never had machines before that could make decisions independently of their human masters. Previously, machines only did what we decided they should do. In some sense, machines in the past were only ever our servants. But we'll soon have machines that make many of their own decisions. Indeed, anyone who owns a Tesla car already has one such machine.

Such autonomy introduces some very difficult new ethical questions. Who is accountable for the actions of an autonomous AI? What limits should be placed on an autonomous AI? And what happens if an autonomous AI harms or kills a person, purposefully or accidentally?

THE RUBBER HITS THE ROAD

The development of self-driving[1] or autonomous cars is one place where we have seen some of the most detailed discussions around AI and ethics. This is unsurprising. A self-driving car is, in fact, a robot. And we don't build many other robots that can travel at over 150 kilometres per hour.

At present, there are approximately 3 million robots working in factories around the world, and another 30 million or so in people's homes. The total number of robots on the planet is thus slightly higher than the population of Australia. But robots will shortly outnumber all humans, and many of these will be self-driving cars.

A car might not look like a robot, but it is. A robot is a machine that can sense, reason and act. A self-driving car can sense the road and other users on the road. It then reasons about where the car needs to go. And finally it acts, following the road and avoiding obstacles. You can simply sit in the car and say, 'Take me home.' And the car will autonomously do the rest.

Self-driving cars will have profound consequences for our lives. One of the most obvious is improved safety. One million people around the world die in road traffic accidents every year. Over a thousand of these deaths are in Australia. This is a billion-dollar problem in Australia alone, as each fatal accident costs around $1 million to clear up. And that's not counting the human cost.

Globally, road traffic accidents are one of the top ten major causes of death. In Australia, if you survive past your first birthday, it is the leading cause of death for the rest of your childhood. And almost all road traffic accidents are caused by human error, not mechanical failure. When we eliminate human drivers from the equation, cars will be much safer.

Computers aren't going to drive tired, or drunk. They won't text while driving, or make any of the other mistakes humans make. They're going to be laser-focused on driving. They'll calculate stopping distances exactly. They'll be looking in all directions simultaneously. They'll devote their vast computational resources to ensuring that no accidents occur.

In 2050, we'll look back at 2000 and marvel at how things were. It'll seem like the Wild West, with people driving cars and having accidents all the time. Often, we underestimate the rate of change. Wind back the clock to 1950 and the roads of Sydney looked much as they do today: full of cars, buses and the odd tram. But go back to 1900 and things were very different. It was mostly horses and carts. The world can become almost unrecognisable in less than 50 years.

By 2050, most of us won't drive. Indeed, most of us won't be able to drive. Young people won't have bothered to learn to drive.

They will go everywhere in self-driving Ubers, which will be as cheap as buses.[2] And old people like you and me will go to renew our driving licences only to discover we haven't had enough practice recently. Driving will be one of those activities, like subtraction or map reading, that humans used to do but that are now done almost exclusively by computers.

Driving in 2050 will be much like horseriding today. It used to be that most people could ride a horse – indeed, it was one of the major forms of transport. Horseriding has now, of course, become a pastime of the rich. Driving a car will go the same way by 2050. It will be an expensive hobby, reserved for special places like race-tracks and tourist destinations.

THE UPSIDE

Self-driving cars won't just make driving safer – they'll have a host of other positive benefits. There are multiple groups who will profit – such as the elderly. My father recently and wisely decided he was no longer able to drive. But once we have self-driving cars, he'll get back that mobility he has lost. Another group that will profit are those with disabilities, who will be given the mobility the rest of us take for granted. Similarly, autonomous cars will benefit the young. I'm looking forward to no longer being a taxi driver for my daughter – I'll let our family car take over the responsibility of ferrying her around. And it will be far less worrying for me to know that a self-driving car is bringing her home late at night from a party, rather than one of her newly licensed friends.

Few inventions of the twentieth century have affected our lives as much as the automobile. It has shaped the very landscape of our cities. It has shaped where we live and work. It gave us mass production. The multistorey car park. Traffic lights and roundabouts. Cars take us to work during the week. To leisure activities at the weekend. And out to the country in the holidays.

The self-driving car will likely redefine the modern city. No longer will the commute to work in the morning and to home at night be 'wasted' time. We can use this time in our self-driving cars to clear our emails, to video-conference with our colleagues, even to catch up on sleep. Living in the suburbs or the country will become more attractive. This will be a good thing, as many of us will no longer be able to afford the ever-inflating property prices of the city centre.

It's easy to describe the primary effect of self-driving cars: we won't be driving anymore. But the secondary effects may be even more interesting. A continuation of the real estate boom in the country kickstarted by the COVID-19 pandemic? Inner cities as places of entertainment rather than of office work? Cars as offices? Might we stop buying cars all together, and just purchase time on some super-Uber self-driving car-sharing service?

Personally, I hate driving. I find it a waste of time. I can't wait to hand over control to a computer, and get back the hours I now spend driving. My only concern is when I'll be able to do so safely.

THE DOWNSIDE

In May 2016, 40-year-old Joshua Brown became the first person killed by a self-driving car. His Tesla Model S was driving autonomously down a highway near Williston, Florida, when an 18-wheel truck pulling a trailer full of blueberries turned across the car's path. It was the middle of a bright spring day.

Unfortunately, the radar on the Tesla likely confused the high-sided vehicle for an overhead sign. And the cameras likely confused the white trailer for the sky. As a result, the car didn't see the truck, did not brake, and drove into the 53-foot-long refrigerated truck at full speed. Actually, it was at more than full speed. The Model S was driving 9 miles per hour faster than the road's speed limit of 65 miles per hour. You may be surprised to know that Tesla's 'Autopilot' lets you set the car's speed significantly above the speed limit.

As the two vehicles collided, the Tesla passed underneath the truck, with the windscreen of the Tesla hitting the bottom of the trailer. The top of the car was torn off by the force of the collision. The Tesla ran on and into a power pole. Joshua Brown died instantly from blunt force trauma to the head.

By many accounts, Joshua Brown was a technophile, an early adopter of new technologies. But like many of us, he appears to have placed a little too much faith in the capabilities of such new technologies. A month earlier, a video of his Tesla on Autopilot avoiding a collision with another truck caught Elon Musk's attention. Joshua Brown tweeted ecstatically:

@elonmusk noticed my video! With so much testing/driving/talking about it to so many people I'm in 7th heaven!

Joshua Brown's hands were on the wheel for only 25 seconds of the 37 minutes of his final journey. The Autopilot system warned him seven times before the fatal crash to place his hands back on the wheel. And seven times he removed his hands from the wheel. According to the Associated Press, the truck driver involved in the accident reported that Brown was actually watching a *Harry Potter* movie at the time of the crash. The police recovered a portable DVD player from the car.

In fact, Joshua Brown might not have been the first person killed by a self-driving car. Four months earlier, Gao Yaning, aged 23, died when the Tesla Model S that he was in drove into a road sweeper on a highway 300 miles south of Beijing in January 2016. However, that crash caused so much damage that Tesla claimed it could not determine if the Autopilot was engaged or not. There have been several more fatal accidents involving self-driving cars since then.

It was inevitable that a self-driving car would eventually kill an innocent pedestrian or cyclist. I made such a prediction at the end of 2016 in a list of AI trends for the coming year.[3] Sadly, it took just over a year for my prediction to come doubly true. In Tempe, Arizona, in March 2018, a self-driving Uber test vehicle struck and killed Elaine Herzberg, a pedestrian pushing her bicycle across the road.

There were both technical and human reasons for this fatal accident. Uber's self-driving system sensed the woman nearly six seconds before the impact. But the system failed to classify her as a pedestrian. She was crossing at a location without a pedestrian crossing, and the system had been set to ignore jaywalkers as it was giving too many false positives. The software also kept changing its classification of her – was she a vehicle? A bicycle? An unknown object? – preventing the car from braking or steering away.

When the self-driving car finally sounded the alarm to instruct Uber's safety driver to intervene, she had only a fraction of a second in which to react. This is where human factors came into play. It didn't help that when the alarm finally went off, the safety driver was watching an episode of *The Voice* on her mobile phone. She was subsequently charged with homicide and is awaiting trial.

The National Transportation Safety Board investigators looking into the accident were highly critical of Uber. They determined that the Arizona testing program lacked a formal safety plan, full-time safety staff and adequate operating procedures. Uber had also reduced its test drivers from two to one per vehicle just five months before the accident.

On the day of the Uber accident, the 3000 people who were killed in other road traffic accidents – those involving human-driven cars – didn't make headlines around the world. Nor did the thousands of safe kilometres that other self-driving cars drove that day make the headlines. But then neither did the report that an Uber engineer sent to Eric Meyhofer, the leader of Uber's self-driving car project. Less than a week before the accident, the engineer's report warned about serious safety problems at Uber. 'We shouldn't be hitting things every 15,000 miles,' the engineer wrote.

Except it wasn't just a 'thing' that Uber was about to hit in March 2018. It was Elaine Herzberg.

HIGH STAKES

The Uber accident in Arizona brought home to me how much was in play in the development of self-driving cars. Actually, it wasn't the fatal accident itself that made me realise how much was in play, but a spooky incident that took place on the night that news of the accident broke.

I was in a taxi going to a TV studio in Sydney to be interviewed on the national nightly news about the accident when my mobile phone rang. It wasn't a number I recognised, but I picked it up just in case it was the producer of the program I was about to go on. The caller identified himself as the CEO of Volvo Australia.

Uber's self-driving car was developed from Volvo's XC90 sports utility vehicle. Uber had taken Volvo's hardware platform and added its own software. Indeed, Volvo's in-house semi-autonomous emergency braking system, which might have prevented the accident, had been switched off by Uber. What was very relevant to this unexpected phone call was that, in 2015, Volvo's global CEO, Håkan Samuelsson, had declared that his company would accept full liability *whenever* one of its cars was in autonomous mode.

My caller wanted me to understand that Volvo was accepting no liability *whatsoever* for the Arizona accident. It was their car. But Uber had changed the software, so Volvo wanted me to know that it was entirely Uber's responsibility.

I was impressed. How did Volvo know I was about to go on TV to discuss the accident? How did they get my mobile phone number? And, given it was their CEO, and not some random PR person, how much must Volvo care?

Around 75 million new cars are sold around the globe every year. That's two cars sold every second. This adds up to well over a trillion dollars of sales annually. Second-hand cars roughly double that again.

The major car companies are worth significant amounts of money. In August 2021, Toyota had a market capitalisation of around $230 billion, the Volkswagen Group was worth $143 billion, Mercedes-Benz $87 billion, General Motors $70 billion, BMW $59 billion, and Honda a little less again at around $52 billion.

However, there is a new kid on the block that has upset this 100-year-old market. Tesla. In 2019, Tesla manufactured 367,500 cars, triple what the company made in 2017. This was still well behind Toyota, which built over 10 million cars in 2019, some 30 times more. Nevertheless, Tesla is already one of the ten most valuable US companies. In August 2021, it was worth over $670 billion, about triple the market cap of Toyota.[4] In fact, Tesla has about a third of the market cap of the whole automotive sector.

The competition doesn't end with cornering the market in building self-driving cars. The race is also on to capture markets around their deployment. Take ride-sharing services like Uber. Perhaps the only way that Uber can turn a profit is if they eliminate the most expensive part of their business: the human sitting in the driver's seat. In this way, all the revenue can come back to Uber. Not surprisingly, Uber is at the forefront of developing self-driving cars.

In August 2021, Uber had a market cap of around $75 billion, on annual revenues of around $14 billion. In addition, there are competitors like Lyft, which has a market cap of around $15 billion, and the Chinese ride-sharing giant DiDi, which is a private company but worth around the same as Uber.

No one knows who the winner in the race to build and deploy self-driving cars will be. Those in Silicon Valley see it largely as a software problem. The winners could therefore be Tesla, Apple, Uber or Waymo (Alphabet), just as much as they could be the incumbents: Toyota, Volvo and General Motors. Whoever wins, it is going to be a very valuable pie. That just might help explain the unexpected phone call I received from the CEO of Volvo Australia while en route to my TV interview.

HOW SELF-DRIVING CARS DRIVE

To understand some of the ethical issues around self-driving cars, as well as the cause of some of these accidents, it helps to have a bit of an understanding about how self-driving cars actually work. You can't yet buy a Level 5, fully autonomous self-driving car that can drive safely on public roads. But we have a good idea of the technology that will go into such a vehicle.

Level 5 means the car requires no human interaction. The vehicle should be able to steer, accelerate, brake and monitor road conditions without needing the driver to pay any attention to the car's functions whatsoever. Level 5 is the highest level of automation defined by the Society of Automotive Engineers, and has been adopted by the US Department of Transportation.[5]

Companies like Waymo and Tesla have tested cars rated at Level 5 on public roads in multiple states in the United States.

Level 5 vehicles have also been deployed in places such as airports and parking lots. In fact, Australia leads the world in this, with fully autonomous vehicles trundling around many mines and ports.

Looking ahead, you can probably expect to come across autonomous vehicles first in geographically constrained spaces, rather than on the open road. This might be a regulated space like a mine. Or it might be the high-speed lane of the motorway, where only autonomous cars that can safely platoon together are allowed.

As to when Level 5 cars will be widely available for sale, it's hard to say. Elon Musk would like you to believe this time might be measured in months. Others have suggested it might be decades. But no one is suggesting it will be never. It's not too soon, therefore, to start planning for their existence.

First and foremost, self-driving cars are very careful map followers. They have high-precision maps, accurate to the centimetre. And they use GPS and other sensors like LIDAR to locate the car precisely within those maps. But you can't drive by just following a map. You need also to sense the world. You need to identify other vehicles and users of the road. You need to avoid obstacles, and cope with places where the map is wrong. Self-driving cars therefore have a range of different sensors with which to observe the world.

One of the primary sensors in a self-driving car is a camera. Just like with human drivers, vision is a good way to observe the world. Cameras are cheap and so can be scattered around the car. Indeed, unlike a human driver, cameras can provide a continuous 360-degree view of the world. However, they have a number of limitations.

Poor weather conditions such as rain, fog or snow will prevent cameras from seeing obstacles clearly. Additionally, there are often situations, like in low light, where the images from these cameras simply aren't adequate for a computer to decide what to do.

Self-driving cars therefore have a number of other sensors. For example, both short-range and long-range radar sensors are deployed all around the car. Radar systems, unlike cameras, are active sensors and so can work at night, or in weather conditions like rain or fog. Long-range radar sensors facilitate distance control and automatic braking. Short-range radar sensors, on the other hand, are used for actions like blind spot monitoring. For even shorter distances, to enable activities like parking, ultrasonic sensors are used.

One of the most useful sensors is LIDAR. This is similar to radar except it uses light produced by a laser instead of radio waves to illuminate the world around the car. LIDAR measures the distance to objects accurately using time of flight, giving a detailed 360-degree cloud of the obstacles surrounding the car. If you've ever seen one of Google's self-driving cars, LIDAR is the rotating ice-cream bucket on the roof.

LIDAR used to be very expensive, costing tens of thousands of dollars. But prices have been plummeting, and LIDAR now costs in the hundreds of dollars. Indeed, LIDAR is so small and cheap now that you'll find it on the latest iPads and iPhones. But while you'll find LIDAR on almost every self-driving car, you won't find it on a Tesla.

Elon Musk has said, 'LIDAR is a fool's errand. Anyone relying on LIDAR is doomed. Doomed! [They are] expensive sensors that are unnecessary. It's like having a whole bunch of expensive appendices. Like, one appendix is bad, well now you have a whole bunch of them, it's ridiculous, you'll see.'[6]

The funny thing is that no other company trying to build autonomous cars besides Tesla is trying to do it without LIDAR. Not using LIDAR is tying your own hands behind your back. The main reason Tesla seems to have gone down this route is that LIDAR was too expensive to put into a Tesla a few years ago. And Tesla wanted to sell you a car then that, after a software upgrade in a few years' time, would provide Level 5 autonomy. Indeed, Tesla is already selling this future software upgrade.

A LIDAR system would very likely have detected the truck crossing the road that killed Joshua Brown in Florida in 2016. Its laser beam would have reflected off the flat side of the trailer as it crossed the path of the car. This would have clearly indicated that the road ahead was blocked. And, instead of driving on blindly, the car would have braked automatically.

I have no desire to repeat Joshua Brown's mistake. I want any self-driving car I trust with my life to be *literally* laser-focused on any obstacles. I won't therefore be trusting a Level 5 self-driving car that lacks LIDAR. And neither should you.

MAGNIFICENT MACHINES

How can it be that companies building self-driving cars get to choose whether to include a safety device like LIDAR? Why is there such limited government oversight of how the industry develops? How can the public be confident that corners are not

74

being cut in the race to market? Can we learn any lessons from other industries, such as aviation or pharmaceuticals?

Suppose a drug company was testing a new product on the public. And imagine that the company didn't have to get approval from an independent ethics board for the trial. That the company wasn't getting informed consent from the public. That the drug was experimental, and the company was expecting some fatal accidents. And that it had, in fact, already killed several members of the public, including innocent bystanders.

If I told you all that, you would be outraged. You'd tell me that surely there are laws preventing this sort of thing. And you'd be right: there are laws preventing such harm from taking place – in the pharmaceutical industry. But this is essentially what is happening today in the development of self-driving cars. Technology companies are testing their prototype self-driving cars on the public roads. They don't have to get much in the way of ethics approval for the trials. They aren't getting informed consent from the public. And they've already killed several people, including innocent bystanders.

How can there not be a government body with careful oversight of the development of self-driving cars? We could perhaps take a lesson from history. One hundred years ago, we developed a new form of transport. And despite the inherent risks, we made it the safest way to get from A to B, by orders of magnitude. This success story is the story of aviation.

In the early days of the aviation industry, those magnificent flying planes were incredibly dangerous. They fell out of the sky with distressing regularity. In 1908, four months after the very first passenger flight, Thomas Selfridge became the industry's first fatality when the Wright Model A flyer crashed, also seriously injuring Orville Wright himself. But somehow the aviation industry turned this around, so that now you're more likely to be killed in a car going to the airport than during the flight itself.

To do this, we set up independent bodies, such as the Australian Transport Safety Bureau, to investigate the cause of accidents, and we mandated that they share those findings with all industry players. We also set up bodies like the Civil Aviation Safety Authority to license pilots, ground crew, aircraft and airfield operators. In this way, airplane accidents don't get repeated. Aircraft safety is like a ratchet: planes only ever get safer.

Contrast this with the competitive race to develop self-driving cars. There is no sharing of information within the industry. Indeed, the only information that the companies developing self-driving cars share is the information they steal from each other. In August 2020, former Waymo engineer Anthony Levandowski was sentenced to 18 months in prison for stealing details of Waymo's self-driving car, which he took with him when he moved to Uber.

TROLLEY PROBLEMS

The most discussed fatal crash involving a self-driving car is – ironically – one that has never happened. Imagine a self-driving car with two people on board that drives around a corner. Crossing the road a short distance ahead is an old lady. There is no time for the car to stop, so the onboard computer has to make a decision. Does the car run over the old lady, surely killing her? Or does it run off the road and into a brick wall, surely killing both its own occupants?

This moral dilemma is known as a 'trolley problem'. It was dreamed up by the English philosopher Philippa Foot in 1967. It's called a trolley problem since the original formulation imagined a runaway trolley heading down a railway track. Five men are working on this track, and are all certain to die when the trolley reaches them. However, it's possible for you to switch the trolley's path to an alternative spur of track, saving all five lives. Unfortunately, one man is working on this spur, and he will be killed if the switch is made. Do you switch the trolley to save the lives of the five men, sacrificing the man on the spur? In surveys, 90 per cent of people switch the trolley, sacrificing the man on the spur but saving five lives.

There are many variants of the trolley problem that explore people's moral decision-making. What if the trolley goes under a footbridge, and there is a fat man on the footbridge who you can push off and into the path of the trolley? The fat man will surely stop the trolley but be killed by the collision. Should you kill the fat man in order to save those five lives? In surveys, some 90 per cent of people would not. Such premeditated murder seems different to throwing the switch in the first example, even though the outcomes are the same: one person dead instead of five.

Another variant of the trolley problem has five people waiting in hospital for transplants of a heart, lung, liver and kidney. Each person will die if they don't have a transplant in the very near future. A fit young person walks into the hospital. You could harvest this young person's organs and save the lives of five people.

Such cold-blooded murder seems very different to many people, even if the choice is the same: to kill one person to save five lives.

There has been so much discussion about trolley problems like this that philosophers now playfully talk about 'trolleyology', the study of ethical dilemmas. This is an academic discipline that brings together philosophy, behavioural psychology and artificial intelligence.

Trolley problems are not, in one sense, new. Every time you've driven a car, you might have had to face such a trolley problem at any moment. But what is new is that we now have to code the answer for how to behave in such a situation in advance. If you study driving codes around the world, none of them specifies precisely what to do in such a circumstance. You will have a split second to decide. If you survive, and you acted unwisely, you may face manslaughter charges. We haven't had to worry previously about how to codify what to do in such a life-or-death situation.

In June 2017, the German Federal Ministry of Transport and Digital Infrastructure released a report containing Germany's ethical guidelines for self-driving cars.[7] The report wishfully recommended that 'the technology must be designed in such a way that critical situations [like the trolley problem] do not arise in the first place'. But since outlawing trolley problems is not actually possible, the report specifies constraints on how autonomous vehicles should be programmed in such situations:

In the event of unavoidable accident situations, any distinction based on personal features (age, gender, physical or mental constitution) is strictly prohibited. It is also prohibited to offset victims against one another. General programming to reduce the number of personal injuries may be justifiable. Those parties involved in the generation of mobility risks must not sacrifice non-involved parties.

There are three fundamental difficulties with the trolley problem. First, 50 years after its creation, we're still struggling to provide a definitive answer to the problem. This shouldn't be surprising. The trolley problem is a moral dilemma – it wasn't designed to have a definitive solution. Dilemmas are, by their very nature, ambiguous, even unanswerable.

The trolley problem permits us to explore the moral tension between deontological reasoning (where we judge the nature of an action rather than its consequences) and consequentialism (where we judge an action by its consequences). *It is wrong to kill* versus *it is good to save lives*. Somehow I doubt that a bunch of computer programmers are going to solve a moral dilemma like this, which has resisted solution by very smart philosophers over decades of argument.

This tension between deontology and consequentialism is at the heart of most variants of the trolley problem. Yet perhaps the dilemma most troubling implication is not the existence of this tension, but the fact that, depending on how the trolley problem is framed, people have wildly different views about the right course of action. How can we hope to code moral behaviours when humans can't agree?

The ethical guidelines of the German Federal Ministry of Transport and Digital Infrastructure reflect this tension. They try to define the dilemma away, mandating that such situations should not occur, so they do not require resolution. While trolley problems are rare, we cannot simply prohibit them by means of some Teutonic precision.

I can still picture the London street on a bright summer morning in 1983 when I faced my own personal trolley problem. I had recently passed my test and was driving my red Mini to work. A car pulled out of a side street to the left of my path. I had a split second to decide between hitting the car or swerving across the road and onto a pedestrian crossing, where a woman and child were crossing. I am not sure if I wisely chose to hit the car or whether I just froze. But in any case there was an almighty bang. The two cars were badly damaged, but fortunately no one was greatly hurt.

The second issue with the trolley problem is that it wasn't dreamed up as a practical moral question. It certainly wasn't intended to describe a real moral dilemma that might be encountered by an autonomous vehicle. It was invented for an entirely different purpose.

The trolley problem was proposed by Philippa Foot as a less controversial way of discussing the moral issues around abortion. When is it reasonable, for example, to kill an unborn child to save the life of the mother? Fifty years later, we still don't have definitive answers to such questions. Abortion was made legal in the United Kingdom in 1967. But it was only decriminalised in New

South Wales in October 2019, having been part of the criminal code for 119 years. How can we expect AI researchers and automotive engineers to code matters of life and death when our broader society has struggled with these issues for decades?

The third and very practical difficulty with the trolley problem is that it is irrelevant to those who program self-driving cars. I know quite a few people who program self-driving cars. If you ask them about the part of the computer code that decides the trolley problems, they look at you blankly. There is no such code. The top-level control loop of a self-driving car is simply to 'drive on the green road'. Traditionally, programmers of self-driving cars have painted the road in front of the car in green to indicate where there are no obstacles and it is safe to drive.[8] And if there's no green road, the car is programmed simply to brake as hard as possible.

Self-driving cars today, and likely for a very long time in the future, simply don't understand the world well enough to solve trolley problems. Self-driving cars are not in a position to trade off lives here for lives over there. The car's perception and understanding of the world is inadequate to make the subtle distinctions discussed in trolley problems.

A self-driving car is like a nervous human learning to drive a car down the road, and saying out loud: *Don't crash ... Don't crash ... Stay on the green road ... Don't crash ... Stay on the green road ... Oh no, I'm going to crash – I'd better brake ... Brake ... Brake ...* If more people appreciated this simple reality, there would be fewer Joshua Browns putting too much trust in the technology and paying with their lives.

MORAL MACHINES

The Media Lab at the Massachusetts Institute of Technology (MIT) is the 'show pony' of the technology world. Ever since it was founded by the charismatic Nicholas Negroponte in 1985, the Media Lab has been famous for its flashy demos. In 2019, it came under fire for secretly accepting donations from convicted child sex offender Jeffrey Epstein. This was after MIT had officially declared that the disgraced financier was disqualified from being a donor.

Negroponte is perhaps best known for the inspirational but poorly executed One Laptop per Child project, launched at the World Economic Forum at Davos in 2005. Negroponte wanted to put $100 laptops in the hands of hundreds of millions of children across the developing world. The project attracted a lot of publicity, but ultimately failed to deliver.

Actually, that's a good description of many Media Lab projects. Another is the Moral Machine. You can check it out at moralmachine.net. It's a platform for crowdsourcing human perspectives on the moral decisions made by machines, such as self-driving cars. Using an interface not too different from Tinder's, humans can vote on how self-driving cars should behave in a particular trolley problem.

> Continue straight into some road works containing a concrete block and kill the two elderly occupants of the car OR swerve across the road and only kill a young child on a pedestrian crossing. How do you vote?

To date, millions of people from almost every country in the world have voted on over 40 million moral choices at moralmachine.net. The goal of the platform is to collect data to 'provide a quantitative picture of the public's trust in intelligent machines, and of their expectations of how they should behave'.[9] Is it that simple? Can we simply program self-driving cars using data from the aptly named 'moral machine'?

There are many reasons to be sceptical. First, we humans often say one thing but do another. We might say that we want to lose weight, but we might still eat a plate full of delicious cream doughnuts. What we tell the moral machine may bear little resemblance to how we might actually behave in a real-world life-or-death situation. Sitting in front of a computer and swiping left or right is nothing like gripping the steering wheel of a car and taking your life in your own sweaty hands.

The second reason to be sceptical of the moral machine is that even if we say what we might *actually* do, there's a lot we actually do that we shouldn't. We are only human, after all, and our actions are not always the right ones. Just because I ate the cream doughnut from my autonomous self-stocking fridge doesn't mean I want the fridge to order me more cream doughnuts. In fact, I want my autonomous fridge to do the opposite. Please stop ordering more fattening temptation.

The third reason to be sceptical is that there's little rigour behind the experiment. But despite this, the inventors of the moral machine have made some bold claims. For example, they say that people in Western countries are more likely to sacrifice the fat man by pushing him off the footbridge than people from Eastern countries, and that this likely reflects differing attitudes to the value of human life in these different societies.[10] Such claims are problematic, as the people using the moral machine are not demographically balanced. They are a self-selecting group of internet users. Unsurprisingly, they are mostly young, college-educated men. In addition, there's no attempt to ensure that their answers are reasonable. To learn more about the moral machine experiment, I took several of its surveys. Each time, I perversely chose to kill the most people possible. The moral machine never once threw me out.

The fourth and final reason to be sceptical of the moral machine is that moral decisions are not some blurred average of what people tend to do. Moral decisions may be difficult and rare. Morality changes. There are many decisions we used to make that we no longer think are moral. We used to deny women the vote. We used to enslave people. We no longer believe either of those are morally acceptable.

Like the trolley problem upon which it is based, the moral machine has attracted a lot of publicity. It is exactly the sort of project you might expect to come out of the Media Lab. But it's much less clear to me whether the moral machine has actually advanced the challenge of ensuring that autonomous machines act in ethical ways.

KILLER ROBOTS

Self-driving cars are not designed to kill. In fact, they are designed to do the opposite – to save lives. But when things go wrong, they may accidentally kill people. There are, however, other autonomous machines entering our lives which are designed expressly to kill: these are 'lethal autonomous weapons' – or, as the media like to call them, 'killer robots'.

The world faces a critical choice about this application of autonomy on the battlefield. This is *not* due to the growing political movement against lethal autonomous weapons. Thirty jurisdictions have called on the United Nations to ban such weapons pre-emptively. I want to list them by name to recognise their moral leadership: Algeria, Argentina, Austria, Bolivia, Brazil, Chile, China, Colombia, Costa Rica, Cuba, Djibouti, Ecuador, Egypt, El Salvador, Ghana, Guatemala, the Holy See, Iraq, Jordan, Mexico, Morocco, Namibia, Nicaragua, Pakistan, Panama, Peru, the State of Palestine, Uganda, Venezuela and Zimbabwe.

Both the African Union and the European Parliament have come out in support of such a ban. In March 2019, the German foreign minister, Heiko Maas, called for international cooperation on regulating autonomous weapons. And in the same week that Maas called for action, Japan gave its backing to global efforts at the United Nations to regulate the development of lethal autonomous weapons.

At the end of 2018, the United Nations' secretary-general, António Guterres, addressing the General Assembly, offered a stark warning:

> The weaponization of artificial intelligence is a growing concern.
>
> The prospect of weapons that can select and attack a target on their own raises multiple alarms – and could trigger new arms races.

Diminished oversight of weapons has implications for our efforts to contain threats, to prevent escalation and to adhere to international humanitarian and human rights law.

Let's call it as it is. The prospect of machines with the discretion and power to take human life is morally repugnant.

However, it's not this growing political and moral concern that underlines the critical choice we face today around killer robots. Nor is it the growing movement within civil society against such weapons. The Campaign to Stop Killer Robots, for instance, now numbers over 100 non-governmental organisations, such as Human Rights Watch which are vigorously calling for regulation. But it's not the pressure of such NGOs to take action that has brought us to this vital juncture.

Nor is it the growing concern of the public around killer robots. A recent international IPSOS poll[11] showed that opposition to fully autonomous weapons has increased 10 per cent in the last two years as understanding of the issues grows. Six out of every ten people in 26 countries strongly opposed the use of autonomous weapons. In Spain, for example, two-thirds of those polled were strongly opposed, while not even one in five people supported their use. The levels were similar in France, Germany and several other European countries.

No, the reason we face a critical choice today about the future of AI in warfare is that the technology to build autonomous weapons is ready to leave the research lab and be developed and sold by arms manufacturers around the world.

In March 2019, for instance, the Royal Australian Air Force announced a partnership with Boeing to develop an unmanned air combat vehicle, a loyal 'wingman' that would take air combat to the next step of lethality. The project is part of Australia's $35-million Trusted Autonomous Systems program, which aims to deliver trustworthy AI into the Australian military. In the same week, the US Army announced ATLAS, the Advanced Targeting and Lethality Automated System, which will be a robot tank. The US Navy, too, announced that its first fully autonomous ship, *Sea Hunter*, had made a record-breaking voyage from Hawaii to the Californian coast without human intervention.

Unfortunately, the world will be a much worse place if, in a decade, militaries around the planet are regularly using such lethal autonomous weapons systems – also known as LAWS – and if there are no laws regulating LAWS. This, then, is the critical choice we face today. Do we let militaries around the world build such technologies without constraints?

The media like to use the term 'killer robot' rather than a wordier expression such as lethal or fully autonomous weapon. The problem is that 'killer robot' conjures up an image of Hollywood's Terminator. And it is not something like the Terminator that worries me or thousands of my colleagues working in AI. It's the much simpler technologies that we see being developed today.

Take the Predator drone. This is a semi-autonomous weapon. It can fly itself for much of the time. However, there is still a soldier, typically in a container in Nevada, who is in overall control of the drone. And, importantly, it is still a soldier who makes the decision to fire one of its deadly Hellfire missiles. But it is a small technological step to replace that soldier with a computer, and empower that computer to identify, track and target. In fact, it is technically possible today.

At the start of 2020, the Turkish military deployed the Kargu quad-copter drone that was developed by the Turkish arms company STM. The drone is believed to be able to autonomously swarm, identify targets using computer-vision and face-recognition algorithms, and destroy those targets on the ground by means of a kamikaze attack.

Once such autonomous weapons are operational, there will be an arms race to develop more and more sophisticated versions. Indeed, we can already see the beginnings of this arms race. In every theatre of war – in the air, on land, and on and under the sea – there are prototypes of autonomous weapons being developed. This will be a terrible transformation of warfare.

But it is not inevitable. In fact, we get to choose whether we go down this road. For several years now, I and thousands of my colleagues, other researchers in the artificial intelligence and robotics research communities, have been warning of these dangerous developments. We've been joined by founders of AI and robotics companies, Nobel laureates, church leaders, politicians and many members of the public.

Strategically, autonomous weapons are a military dream. They let a military scale their operations unhindered by manpower constraints. One programmer can command hundreds of autonomous weapons. This will industrialise warfare. Autonomous weapons will greatly increase strategic options. They will take humans out of harm's way, making the riskiest of missions more feasible. You could call it War 3.0.

There are many reasons, however, why the military's dream of lethal autonomous weapons systems will turn into a nightmare. First and foremost, there is a strong moral argument against killer robots. We give up an essential part of our humanity if we hand over military decisions to a machine. Machines have no emotions, compassion or empathy. How, then, can they be fit to decide who lives and who dies?

Beyond the moral arguments, there are many technical and legal reasons to be concerned about killer robots. In the view of my colleague Stuart Russell, co-author of the definitive textbook on AI, one of the strongest arguments is that they will revolutionise warfare. Autonomous weapons will be weapons of immense and targeted destruction. Previously, if you wanted to do harm, you had to have an army of soldiers to wage war. You had to persuade this army to follow your orders. You had to train them, feed them and pay them. Now a single programmer could control hundreds of weapons.

Lethal autonomous weapons are more troubling, in some respects, than nuclear weapons. To build a nuclear bomb requires technical expertise – you need skilled physicists and engineers. You need the resources of a nation-state, and access to fissile material. Nuclear weapons have not, as a result, proliferated greatly. Autonomous weapons require none of this. You'll just need a 3D printer and some sophisticated code that is easily copied.

If you're not yet convinced that autonomous weapons might pose a bigger threat than nuclear weapons, then I have some more bad news for you. Russia has announced plans to build Poseidon, an autonomous nuclear-powered and nuclear-tipped underwater submarine.[12] Can you think of anything more horrifying than an algorithm that can decide to start a *nuclear* war?

But even non-nuclear autonomous weapons will be a dreadful scourge. Imagine how terrifying it will be to be chased by a swarm of autonomous kamikaze drones. Autonomous weapons will likely fall into the hands of terrorists and rogue states, who will have no qualms about turning them on innocent civilians. They will be an ideal weapon with which to suppress a population. Unlike humans, they will not hesitate to commit atrocities, even genocide.

LAWS BANNING LAWS

You may be surprised to hear this, but not everyone is on board with the idea that the world would be a better place with a ban on killer robots. 'Robots will be better at war than humans,' they say. 'They will follow their instructions to the letter. Let robot fight robot and keep humans out of it.'

Such arguments don't stand up to much scrutiny, in my view. Nor do they stand up to the scrutiny of many of my colleagues in AI and robotics. Here are the five main objections I hear to banning killer robots – and why they're misguided.

Objection 1: Robots will be more effective than humans.
They'll be more efficient for sure. They won't need to sleep. They

won't need time to rest and recover. They won't need long training programs. They won't mind extreme cold or heat. All in all, they'll make ideal soldiers.

But they won't be more effective. The leaked 'Drone Papers', published by the *Intercept* in 2015, recorded that nearly nine out of ten people killed by drone strikes weren't the intended target.[13] And this is when there was still a human in the loop, making the final life-or-death decision. The statistics will be even worse when we replace that human with a computer.

Killer robots will also be more efficient at killing us. Terrorists and rogue nations are sure to use them against us. It's clear that if the weapons not banned, then there will be an arms race. It is not over-blown to suggest that this will be the next great revolution in warfare, after the invention of gunpowder and nuclear bombs. The history of warfare is largely one of who can more efficiently kill the other side. This has typically not been a good thing for humankind.

Objection 2: Robots will be more ethical.

This is perhaps the most interesting argument. Indeed, there are even a few who claim that this moral argument *requires* us to develop autonomous weapons. In the terror of battle, humans have committed many atrocities. And robots can be built to follow pre-cise rules. However, as this book shows, it's fanciful to imagine we know how to build ethical robots. AI researchers have only just started to worry about how you could program a robot to behave ethically. It will take us many decades to work out how to deploy AI responsibly, especially in a high-stakes setting like the battlefield.

And even if we could, there's no computer we know of that can't be hacked to behave in ways that we don't desire. Robots today cannot make the distinctions that the international rules of war require: to distinguish between combatant and civilian, to act proportionally and so on. Robot warfare is likely to be a lot more unpleasant than the wars we fight today.

Objection 3: Robots can just fight robots.

Replacing humans with robots in a dangerous place like the battlefield might seem like a good idea. However, it's fanciful to suppose that we could just have robots fighting robots. Wars are fought in our towns and cities, and civilians are often caught in the crossfire, as we have sadly witnessed recently in Syria and elsewhere. There's not some separate part of the world called 'the battlefield' where there are only robots.

Also, our opponents today are typically terrorists and rogue states. They are not going to sign up to a contest between robots. Indeed, there's an argument that the terror unleashed remotely by drones has likely aggravated the many conflicts in which we find ourselves enmeshed currently. To resolve some of these difficult situations, we have to put not robots but boots on the ground.

Objection 4: Such robots already exist, and we need them.

There are some autonomous weapons deployed by militaries today – weapons systems like the Phalanx CIWS (close-in weapon system), which sits on many US, British and Australian naval ships. Or Israel's Harpy weapon, a kamikaze drone that loiters for up to six hours over a battlefield and uses its anti-radar homing system to take out surface-to-air missile systems on the ground.

I am perfectly happy to concede that a technology like the autonomous Phalanx system is a good thing. But the Phalanx is a *defensive* system, and you don't have time to get a human decision when defending yourself against an incoming supersonic missile. I and other AI researchers have called for *offensive* autonomous systems to be banned, especially those that target humans. An example of the latter is the autonomous Kargu drone, currently active on the Turkish–Syrian border. This uses the same facial-recognition algorithms as in your smartphone, with all of their errors, to identify and target people on the ground.

There's no reason why we can't ban a weapons system that already exists. Indeed, most bans, like those for chemical weapons or cluster munitions, have been for weapons systems that not only exist but have been used in war.

Objection 5: Weapon bans don't work.

History would contradict the argument that weapon bans don't work. The 1998 UN Protocol on Blinding Lasers has resulted in lasers that are designed to blind combatants permanently being kept off the battlefield. If you go to Syria today – or to any other war zones – you won't find this weapon. And not a single arms company anywhere in the world will sell you one. You can't uninvent the technology that supports blinding lasers, but there's enough stigma associated with them that arms companies have stayed away.

I hope a similar stigma will be associated with autonomous weapons. It's not possible to uninvent the technology, but we can put enough stigma in place that robots aren't weaponised. Even a partially effective ban would likely be worth having. Anti-personnel mines still exist today despite the 1997 Ottawa Treaty. But 40 million such mines have been destroyed. This has made the world a safer place, and resulted in many fewer children losing a limb or their life.

*

AI and robotics can be used for many great purposes. Much of the same technology will be needed in autonomous cars as in autonomous drones. And autonomous cars are predicted to save 30,000 lives on the roads of the United States alone every year. They will make our roads, factories, mines and ports safer and more efficient. They will make our lives healthier, wealthier and happier. In the military setting, there are many good uses of AI. Robots can be used to clear minefields, take supplies through dangerous routes and shift mountains of signal intelligence. But they shouldn't be used to kill.

We stand at a crossroads on this issue. I believe it needs to be seen as morally unacceptable for machines to decide who lives and who dies. If the nations of the world agree, we may be able to save ourselves and our children from this terrible future.

In July 2015, I helped organise an open letter to the United Nations calling for action that was signed by thousands of my colleagues. The letter was released at the start of the main international AI conference. Sadly, the concerns we raised in this letter have yet to be addressed. Indeed, they have only become more urgent. Here is the text of our open letter:

AUTONOMOUS WEAPONS: AN OPEN LETTER FROM AI & ROBOTICS RESEARCHERS

Autonomous weapons select and engage targets without human intervention. They might include, for example, armed quadcopters that can search for and eliminate people meeting certain pre-defined criteria, but do not include cruise missiles or remotely piloted drones for which humans make all targeting decisions. Artificial Intelligence (AI) technology has reached a point where the deployment of such systems is – practically if not legally – feasible within years, not decades, and the stakes are high: autonomous weapons have been described as the third revolution in warfare, after gunpowder and nuclear arms.

Many arguments have been made for and against autonomous weapons, for example that replacing human soldiers by machines is good by reducing casualties for the owner but bad by thereby lowering the threshold for going to battle. The key question for humanity today is whether to start a global AI arms race or to prevent it from starting. If any major military power pushes ahead with AI weapon development, a global arms race is virtually inevitable, and the endpoint of this technological trajectory is obvious: autonomous weapons will become the Kalashnikovs of tomorrow.

Unlike nuclear weapons, they require no costly or hard-to-obtain raw materials, so they will become ubiquitous and cheap for all significant military powers to mass-produce. It will only be a matter of time until they appear on the black market and in the hands of terrorists, dictators wishing to better control their populace, warlords wishing to perpetrate ethnic cleansing etc. Autonomous weapons are ideal for tasks such as assassinations, destabilising nations, subduing populations and selectively killing a particular ethnic group. We therefore believe that a military AI arms race would not be beneficial for humanity. There are many ways in which AI can make battlefields safer for humans, especially civilians, without creating new tools for killing people.

Just as most chemists and biologists have no interest in building chemical or biological weapons, most AI researchers have no interest in building AI weapons – and do not want others to tarnish their field by doing so, potentially creating a major public backlash against AI that curtails its future societal benefits. Indeed, chemists and biologists have broadly supported international agreements that have successfully prohibited chemical and biological weapons, just as most physicists supported the treaties banning space-based nuclear weapons and blinding laser weapons.

In summary, we believe that AI has great potential to benefit humanity in many ways, and that the goal of the field should be to do so. Starting a military AI arms race is a bad idea, and should be prevented by a ban on offensive autonomous weapons beyond meaningful human control.[14]

In 2020, five years after we wrote this letter, the United Nations is still discussing the idea of regulating killer robots. And the military AI arms race that we warned about has clearly begun.

THE RULES OF WAR

If the United Nations fails to prohibit killer robots in the near future, we will have to work out how to build robots that will follow the rules of war. From the outside, war might appear to be a rather lawless activity. A lot of people get killed in war, and killing people is generally not permitted in peacetime. But there are internationally agreed rules for fighting war. And these rules apply to robots as much as to people.

The rules of war distinguish between *jus ad bellum* – one's right to go to war – and *jus in bello*, one's rights while at war. To put it in plainer language, the rules of war distinguish between the conditions under which states may resort to war and, once states are legally at war, the way they conduct warfare. The two concepts are deliberately independent of each other. *Jus ad bellum* requires, for example, that war must be fought for a just cause, such as to save life or protect human rights. It also requires that war must be defensive and not aggressive, and that it must be declared by a competent authority such as a government. For the present, it is unlikely that machines are going to be declaring war by themselves. It is perhaps reasonable to suppose, therefore, that humans are still going to be the ones taking us to war. So I'll put aside for now concerns about killer robots accidentally starting a war, and focus instead on *jus in bello*.

The rules governing the conduct of war seek to minimise suffering, and to protect all victims of armed conflict, especially non-combatants. The rules apply to both sides, irrespective of the reasons for the conflict or the justness of the causes for which they are fighting. If it were otherwise, the laws would be pretty useless, as each party would undoubtably claim to be the victim of aggression.

There are four main principles of *jus in bello*. We begin with the principle of humanity, which also goes under the name of the 'Martens Clause'. This was introduced in the preamble to the 1899 Hague Convention by Friedrich Martens, a Russian delegate. It requires war to be fought according to the laws of humanity, and the dictates of the public conscience.

The Martens Clause is a somewhat vague principle, a catch-all that outlaws behaviours and weapons that the public might find repugnant. How, for instance, do we determine precisely the public conscience? The Martens Clause is often interpreted to prefer, for example, capturing an enemy over wounding them, and wounding over killing, and to prohibit weapons that cause excessive injury or pain.

The second principle of *jus in bello* is that of distinction. You must distinguish between the civilian population and combatants, and between civilian objects and military objectives. The only legitimate target is a military objective. It requires defenders to avoid placing military personnel or matériel in or near civilian objects, and attackers to use only those methods of assault that are discriminating in effect.

The third principle of *jus in bello* is that of proportionality. This prohibits attacks against military objectives which are expected to cause incidental loss of civilian life, injury to civilians or damage to civilian objects which would be excessive compared to the expected military advantage from that attack. This principle requires attackers to take precautions to minimise collateral damage, and to choose, where possible, objectives likely to cause the least danger to civilians and civilian objects.

The fourth and final principle of *jus in bello* is that of military necessity. This limits armed force to those actions that have legitimate military objectives. This means avoiding inflicting gratuitous injury on the enemy. The principle of necessity overlaps in part with the Martens Clause. Both take account of humanitarian concerns around the wounding of soldiers. And both prohibit weapons that cause unnecessary suffering.

In my view, lethal autonomous weapons today fail to uphold all four principles of *jus in bello*, the conduct of war. Consider, for example, the Martens Clause. The majority of the public are against the idea of lethal autonomous weapons. Indeed, as the UN secretary-general clearly said, many of us find them morally repugnant. It seems therefore that lethal autonomous weapons conflict directly with the Martens Clause.

The other three principles are also violated by lethal autonomous weapons. For instance, we don't know how to build weapons that can adequately distinguish between combatant and civilian. The Kargu drone deployed on the Turkish–Syrian border uses facial-recognition technology to identify targets. And yet we know that, in the wild, such facial-recognition software can be incredibly inaccurate.[15] It is hard, then, to imagine how the Kargu drone upholds the principle of distinction.

What's more, we cannot yet build autonomous systems that respect the principles of proportionality and necessity. We can build autonomous systems like self-driving cars that perceive the world well enough not to cause an accident. But we cannot build systems that make subtle judgements about the expected damage a particular weapon will inflict. Or about the humanitarian trade-offs between a variety of different targets.

I am willing to concede that some of the principles of *jus in bello*, like that of distinction, may be achieved by AI systems at some point in the future. In a couple of decades, for example, machines may be able to distinguish adequately between combatants and civilians. Indeed, there are arguments that machines may one day be better at upholding the principle of distinction than humans. After all, machines can have more sensors, faster sensors, sensors that work on wavelengths of light that humans cannot see, even active sensors, like radar and LIDAR, which work in conditions that defeat passive sensors like our eyes and ears. It is plausible, then, that killer robots will one day perceive the world better than we humans can.

However, there are other principles, such as the Martens Clause, that it is hard to imagine machines will ever be able to uphold. How will a machine understand repugnance? How can a machine determine the public conscience? Similar concerns arise around the principles of proportionality and necessity. Could a machine ever adequately understand the humanitarian concerns that a military commander considers when some insurgents are hiding near a hospital?

In February 2020, the US Department of Defense officially announced the adoption of a series of ethical principles for the use of artificial intelligence within the military.[16] The principles emerged from over a year of consultation with AI experts, industry, government, academia and the American public. They apply to both combat and non-combat situations. The ethical principles boldly promise AI that is responsible, equitable, traceable, reliable and governable.

US Department of Defense's ethical principles for the use of AI

1. Responsible. [Department of Defense] personnel will exercise appropriate levels of judgment and care, while remaining responsible for the development, deployment, and use of AI capabilities.

2. Equitable. The Department will take deliberate steps to minimize unintended bias in AI capabilities.

3. Traceable. The Department's AI capabilities will be developed and deployed such that relevant personnel possess an appropriate understanding of the technology, development processes, and operational methods applicable to AI capabilities, including with transparent and auditable methodologies, data sources, and design procedure and documentation.

4. Reliable. The Department's AI capabilities will have explicit, well-defined uses, and the safety, security, and effectiveness of such capabilities will be subject to testing and assurance within those defined uses across their entire life-cycles.

5. Governable. The Department will design and engineer AI capabilities to fulfil their intended functions while possessing

the ability to detect and avoid unintended consequences, and the ability to disengage or deactivate deployed systems that demonstrate unintended behaviour.

It is hard to disagree with many of these desires of the US Department of Defense. Who would want an *unreliable* autonomous tank that was sometimes responsible for friendly-fire casualties? Or a kamikaze drone that was *biased* against Black people, causing more accidental civilian deaths in Black populations than in white populations? As with other announcements of ethical principles for AI systems, two fundamental questions remain: could we achieve such laudable aims? And, if so, how would we go about it?

HUMANS V. MACHINES

LIFE 1.0

If we are going to build artificial intelligence, and especially if we are going to build machines with autonomy such as self-driving cars or killer robots, it seems hard to avoid the necessity that such machines behave ethically. I therefore want to spend some time considering *how* we might do that. How can machines possibly be programmed to do the right thing?

One place to start is with the similarities between machine and human intelligence, since humans already can, and often do, behave ethically. Could we perhaps reproduce in machines the ethical decision-making that humans perform? Or are human intelligence and machine intelligence so fundamentally different that we cannot replicate this?

One important difference between humans and machines is that we're alive and machines aren't. But what does it mean to be alive? Even though our experience of being alive is central to our existence, there is surprisingly little consensus among scientists, philosophers or others interested in the question of precisely what life is, and what it means to be alive.

Even defining what is or isn't life is a major challenge. Popular definitions of living systems include a wide variety of characteristics: they are systems that maintain some sort of equilibrium, have a life cycle, undergo metabolism, can grow and adapt to their environment, can respond to stimuli, and can reproduce and evolve.

You might be able to tell from this long and rather motley collection of characteristics that biologists don't really know what life is. They've simply added more and more features of life to their definition until everything else is excluded. Indeed, this list currently excludes one of 2020's greatest foes, the humble virus.

Machine intelligence already has, or is likely to have, many of these characteristics of life. For example, AI programs already adapt to their environment. The self-driving car will swerve around a child that runs onto the road. A smart thermostat will change the room temperature according to the behaviours it has learned of the occupants of the house. Other AI programs can *evolve* to perform better. In fact, there's a branch of AI called genetic programming that has evolution at its core.

The characteristics of life are important to ethical behaviour. The fact that we have a life cycle and will eventually be dead is reflected in the value ethics places on preserving life. The fact that we respond to stimuli and can experience pain is reflected in the value ethics places on avoiding suffering.

The ancient Greek philosopher Aristotle argued that ethics helps us to live a good life. And a necessary part of living a good life is to be alive. Without life, therefore, we would have no need for any ethics.

But whether in some distant future we build machines sufficiently complex and adaptive that we might think of them being alive is immaterial to how such machines need to behave. We cannot have a world in which living creatures are harmed because of the careless design of artificially intelligent machines, be they animate or inanimate.

THE DEMON IN THE MACHINE

One very important feature that machines appear to lack is free will. And free will is central to our ethical behaviours. It is precisely because we have free will that we get to worry about making the *right* ethical choices. Indeed, if we didn't have free will, there would be no choices, let alone ethical choices, to make.

There is, of course, a huge assumption in my argument: that humans have free will. Science has so far failed to address this assumption in a meaningful way. Free will cannot be found anywhere in the laws of physics, chemistry or biology. Given a particular state of the world, physics tells us how to compute the next state. Even in the weirdest quantum mechanical systems, you predict the next state merely by tossing a coin. There is no place where the human mind gets to choose which outcome happens.

But it certainly *seems* like we humans have free will. I could choose, for instance, to end this paragraph right here.

See, I have free will – I ended the paragraph right there. And I imagine that you think the same about your free will. You can put this book down right now. No, please don't.

But machines – surely they're much simpler and just follow the laws of physics? There's no need – indeed, no place – for free will in describing their operation and in understanding their behaviours. Computers are deterministic machines that simply follow the instructions in their code.

One problem with this argument is that machines are becoming more complex by the day. The memory capacity of the human brain is around a petabyte, in the same ballpark as the World Wide Web.[1]

Once computers are more complex than human brains, it will become harder to claim that free will only emerges out of the sophistication of human brains.

Another problem with this argument is that complexity also arises out of the interaction with the real world. Machines are embedded in the physical world. There are lots of examples of how rich, complex behaviours can emerge in such a situation. A butterfly flaps its wings and alters the path of a tornado.

We might therefore look for other features that machines lack, such as consciousness. In fact, consciousness seems closely connected to free will. Is it not precisely because you are conscious of the different ethical choices ahead that you can exercise free will?

A lack of consciousness could actually be a barrier to machines matching human intelligence. In 1949, Sir Geoffrey Jefferson eloquently put this argument forward in the ninth Lister Oration:

Not until a machine can write a sonnet or compose a concerto because of thoughts and emotions felt, and not by the chance fall of symbols, could we agree that machine equals brain – that is, not only write it but know that it had written it. No mechanism could feel (and not merely artificially signal, an easy contrivance) pleasure at its successes, grief when its valves fuse, be warmed by flattery, be made miserable by its mistakes, be charmed by sex, be angry or depressed when it cannot get what it wants.[2]

Of course, consciousness in humans is also poorly understood by science. There is, however, hope that this may change in the near future. Neurobiologists are making increasingly optimistic noises that they are beginning to understand the biology of consciousness. Indeed, artificial intelligence may throw some light on this question.

It is not clear whether computers, on the other hand, will ever develop some sort of consciousness. Perhaps it is a uniquely biological phenomenon? In truth, we might prefer that machines are not able to gain consciousness. Once machines are conscious, we may have ethical obligations to them in how we treat them. For instance, can we now turn them off?

In any case, since we understand so little today about consciousness, it is not at all clear to me that consciousness is necessarily a fundamental difference between artificial and human intelligence. Perhaps we can have intelligence without consciousness? Or perhaps intelligence is something that emerges, given sufficient intelligence, in machine or in biology? We certainly cannot tolerate a world in which unethical behaviours are harming many conscious entities merely because the machines carrying out those beahviours are not conscious.

EMOTIONS

Another feature machines lack is emotions. We can say this with certainty, as emotions have a significant chemical component, and computers are not chemical devices. Evolution, on the other hand, has given us rich emotional lives. Emotions must therefore play an important role in our fitness to survive.

Emotions are physical and psychological changes that influence our behaviour. This can provide a short circuit to speed up human decision-making. There are six basic emotions: anger, disgust, fear, happiness, sadness and surprise. Each helps us to respond to a given situation. Anger strengthens our ability to stand and fight. Disgust drives us away from a potential harm. Fear helps us run away from danger. Happiness reinforces the positive behaviours that brought this satisfaction. Sadness discourages the negative behaviours that brought gloom into our lives. And surprise motivates us to discover and learn more from the world.

We may like to think that our ethical decision-making is the result of conscious thought and absolute values, but the reality is that we are often driven by emotions. Inner-looking and negative emotions, such as guilt, embarrassment and shame, may motivate us to act ethically. And outward-looking and more positive emotions, such as empathy and sympathy, can prompt us to help others.

Daniel Kahneman, a behavioural economist from Princeton, was awarded the Nobel Prize in Economics in 2002 for his work questioning the conscious rationality of human decision-making.[3] He has argued that 98 per cent of our thinking is 'System 1', which is fast, unconscious, automatic and effortless. The remaining 2 per cent of our thinking is 'System 2', which is deliberate, conscious, effortful and rational.

If this is the case, then can we understand and replicate the ethical choices in human decision-making without understanding and perhaps replicating the subconscious and emotional basis behind most of it? Alternatively, might AI give us the opportunity to replace evolution's outdated, irrational, unconscious and emotional kludges with something more precise and rational?

PAIN AND SUFFERING

An important aspect of the human experience is pain and suffering. Life begins and sometimes ends in pain. And, sadly, some of the in-between involves pain and suffering too. This is not something that machines experience.

Pain begins with an electrical signal travelling along a nerve. But pain has a chemical basis, involving a complicated process of neurotransmitters, the chemical messengers that signal pain, along with endorphins, the natural opiates released in response to pain. Computers have none of this biochemical complexity.

It would actually be useful to build robots that experience pain (or its electronic equivalent). Pain is an important mechanism for avoiding harm. We remove our hand from the fire in response to the pain we feel, not because we stop to reason about the harm the heat will cause our body. It's a simple and rapid response to pain. Having robots experience 'pain' so that they respond to dangerous situations in a similar fashion might help prevent them from coming to harm. We could program a register that records their level of pain, and have the robot act to keep this as low as possible. But this artificial pain doesn't seem to carry the same moral weight as the real pain that humans (and animals) suffer.

Suppose for a moment that we could give robots something approaching real pain. Would it be moral to do so? If they really could suffer, we would have to worry about their suffering. This would greatly limit their usefulness. We might not be able to have them do all our dirty and dangerous work.

And if we gave robots pain, we might not stop there. We might also decide to give them fear, which often precedes pain and also prevents injury. But why stop with just these emotions? Might it not be useful to give computers the full spectrum of human emotions, and let them be happy, sad, angry and surprised too?

If computers had all these human emotions, might they fall in love, create music that brings us to tears and write poems that speak to the joys and sorrows of life? Perhaps – but they might also become anxious, get angry, act selfishly and start wars. Giving computers emotions could open a whole can of ethical worms.

AI = ALIEN INTELLIGENCE

With or without emotions, artificial intelligence might be very different to human intelligence. When I talk about AI in public, I remind people to focus on the word *artificial* as much as on the other word, *intelligence*. And AI might be very *artificial* indeed. It could be a remarkably distinct form of intelligence to the natural intelligence that we have.

Flight is a good analogy. Artificial flight – that is, the flight that humans engineered with our intelligence – is very different to natural flight, or the flight that nature discovered through evolution. We have built airplanes that can circle the globe, fly faster than the speed of sound and carry tonnes of cargo. If we'd only ever tried to re-create natural flight, I imagine we'd still be on the ground, flapping our wings and watching the birds high above us.

We came at the problem of flight from a completely different angle to nature: with a fixed wing and a powerful engine, not with moving wings, feathers and muscles. Both natural and artificial flight depend on the same Navier–Stokes equations of fluid dynamics. But they are different solutions to the problem. Nature doesn't necessarily find the easiest or the best solution to any problem.

Artificial intelligence might similarly be a very different solution to the problem of intelligence compared to natural intelligence. We know already that intelligence comes in different forms, as humans are not the only intelligent life on the planet. Nature has found several other forms of intelligence.

The amazing octopus, perhaps the smartest of all invertebrates, is a good example of a different intelligence. Octopuses can open a screw-cap jar. They can use tools, which is often a measure of intelligence. Octopuses cooperate and communicate when hunting together, skills that are also often associated with intelligence. People who work with octopuses claim that they can recognise faces and remember people. These humans will often ascribe personality to their captive charges. Who cannot be moved by the incredible scenes in the documentary *My Octopus Teacher*?

Octopuses appear to dislike being held in captivity. They are famous escapologists. And who was not amazed by Paul the Octopus's ability to predict the result of football matches at the 2010 World Cup?

Unlike most other invertebrates, octopuses have been given protection from scientific testing in several countries. In the United Kingdom, for example, the common octopus is the only invertebrate protected by the *Animals (Scientific Procedures) Act 1986*. This limits the use of octopuses in any experimental or other scientific procedure which may cause pain, suffering, distress or lasting harm.

So what do we know about the differences between octopus and human intelligence? All animals are related, humans included: we just have to look far enough back in our evolutionary tree to find a common ancestor. In the case of humans and octopuses, that is a long, long time ago. Octopuses diverged from humans around 600 million years ago. To put that in perspective, it is long before the dinosaurs were in charge of the planet. At the time in our evolutionary history when humans separated from octopuses, the most complex animals in existence only had a few neurons. Therefore, whatever intelligence octopuses have today evolved entirely separately to ours.

And how differently the octopus evolved! The giant Pacific octopus, for example, has three hearts, nine brains and blue blood, and can change the colour of its skin in the blink of an eye. Though they are underwater creatures, octopuses don't need to blink. Each leg of an octopus can sense and think independently of the other seven. In some way, the octopus is nine brains working as one.

112

The octopus is perhaps the closest thing, then, that we have on Earth to an alien intelligence. And that may be the best way to imagine AI – as an alien (rather than artificial) intelligence. And the limited AI we have built today has all the appearance of an alien intelligence, as it bears little resemblance to human intelligence.

Human perception, for example, is incredibly robust. You can rotate an image and not change how we see the world. Indeed, if you wear glasses that completely invert the world, your brain soon compensates and turns the world back up the correct way. Computer vision is, by comparison, very brittle. You can rotate an image a few degrees and confuse any computer-vision system. In fact, you can change a single pixel and fool the algorithms. The computer stops labelling the image as a bus and starts labelling it as a banana. This should be very worrying to those building self-driving cars. Computers see the world in a very different way to us humans.

As another example, computers understand language in a much more statistical way than humans. Give Google Translate a sentence to translate into French like 'He is pregnant'. You'll get back 'Il est enceinte'. The computer knows statistically that 'He' translates as 'Il', 'is' as 'est' and 'pregnant' as 'enceinte'. And this translation is grammatically correct. But the computer doesn't have the mental model and understanding that you or I have. The computer doesn't construct a mental picture of a pregnant man. Or start wondering about seahorses, whose males and not females become pregnant. Or whether the human race would break down if men had to give birth. The computer's understanding of language is statistical, not semantical.

It is for these reasons that it may be best to think of AI as alien intelligence. There is no a priori reason that artificial intelligence has to be the same as human intelligence.

ROBOT RIGHTS

Driven by these differences between humans and machines, we have not as yet given machines any rights. I can go down to my laboratory and torture my robots as much as I like. Overload their capacitors. Remove their gears one by one. The authorities won't care one jot.

Humans, on the other hand, have many rights. The right to life, liberty and security of person. The right to equality before the law. The right to freedom of movement. The right to leave any country, including that of your birth.[4] As you may have worked out, I'm paraphrasing from the Universal Declaration of Human Rights.

Of course, this doesn't mean humans don't care at all about robots. In Japan, when Sony stopped producing the cute AIBO robot dog, an electronics company started holding Buddhist funerals in a historic temple to farewell the mechanical pets when they could no longer be repaired.

Boston Dynamics, a robot company that spun out of MIT, released some impressive videos on YouTube of Atlas, its humanoid robot, doing backflips and other amazing feats.[5] Videos of BigDog and Spot, its rather sinister-looking four-legged robots, inspired a terrifying episode of *Black Mirror*. However, the responses to images of the Boston Dynamics robots being knocked over and hit repeatedly was telling. CNN ran a story with the headline 'Is It Cruel to Kick a Robot Dog?'.[6] A website appeared with a somewhat tongue-in-cheek campaign to stop robot abuse.[7] And Alphabet, Google's parent company, decided to sell Boston Dynamics to SoftBank shortly after. I suspect that the optics of robots being abused in public contributed to the decision.

Putting aside such sentimental attachments – which we also see with bicycles and other machines that become parts of our lives and yet that are clearly not intelligent – there are few protections given to intelligent machines. Can this continue once they become smarter? We give rights to other animals when we understand that they experience pain and suffering. In general, invertebrates have very simple nervous systems and, as far as we know, limited capability to experience pain. As a consequence, invertebrates have few protections under the laws of most countries. Mammals, on the other hand, have complex nervous systems and enjoy wide protections from being subjected to pain and suffering.

If robots are unable to experience pain and suffering, then it might follow that, no matter how intelligent they become, they are not in need of any rights. We can treat them like any other machine. Toasters have no rights. Robots are often portrayed as our servants. They can take over the dirty, the dull, the difficult and even the dangerous parts of our lives. But the real question is not whether they are our servants, but whether they are merely chattels. Will they ever be more than just our property?

SOPHIA THE PUPPET

In October 2017, an incredibly realistic humanoid robot produced by Hanson Robotics and called Sophia was granted citizenship of

Saudi Arabia. David Hanson, the founder of Hanson Robotics, is very good at attracting publicity for his robot creations. And having Sophia become the world's first robot citizen garnered headlines around the globe.

Sophia is, however, little more than a fancy puppet. You should know that David Hanson was previously an engineer building the realistic animatronic figures produced by Disney. You can find these marvellous animatronic creations at theme parks around the world. But back to Hanson Robotics. Sophia has not much in the way of artificial intelligence. She mostly follows a script that humans have written in advance of any public appearance. Apple's Siri is arguably significantly more intelligent than Sophia.

It's somewhat ironic that Saudi Arabia, a country with a problematic record on human rights, should be the first nation to grant rights to a robot. Especially to a robot that is more or less a puppet. And it is troubling that Sophia was given citizenship when over a third of the population of Saudi Arabia, including non-Muslims and foreign workers, are denied many basic human rights.

Granting rights to robots is a fundamentally flawed idea that is not limited to Saudi Arabia. In 2017, the European Parliament proposed:

> creating a specific legal status for robots in the long run, so that at least the most sophisticated autonomous robots could be established as having the status of electronic persons responsible for making good any damage they may cause, and possibly applying electronic personality to cases where robots make

autonomous decisions or otherwise interact with third parties independently.[8]

This legal personhood for robots would be similar to that already given to corporations, so that we can make corporations liable for their actions. But haven't we been struggling in recent years to hold corporations accountable? Haven't we been troubled by tobacco companies, oil companies and tech companies behaving unethically? It hardly seems a good idea to introduce yet another type of personhood, and to expect this to align with the values of society. If robots don't experience pain or suffering, and we are not therefore obliged to give them rights, is it a good idea to choose to do so?

Giving rights to robots introduces a host of problems. We would have to hold them accountable for their actions. But how could we possibly do that? We give rights to humans (and other animals) to protect their lives, as life is precious and finite. But AI is neither precious nor finite. We can easily duplicate any AI program, and such a program can run forever. What then do we need to protect?

Giving rights to robots places an unnecessary burden on humans. We would then have to spend effort respecting these rights. We might even have to sacrifice some of our own human rights in their favour. In addition, robots may then be unable to do tasks that would have protected the rights of humans. Giving rights to robots might therefore be not just morally unnecessary but morally harmful to people who actually deserve rights.

We can break this problem down into two parts. Can robots have rights? And should robots have rights? The first question asks whether robots (or other AIs) are made of the right stuff to have moral capability. The second question asks whether, given their form, they deserve certain rights.

I've raised doubts about whether robots can have rights. There are some fundamental differences between robots (and other AIs) today and humans, such as consciousness, that might preclude them from behaving ethically. And the jury is out on whether such differences will remain in the future. I've also raised doubts about whether robots (and other AIs) should have rights. Giving rights to robots may impose an unnecessary burden on humans, and even weaken our ability to protect human rights.

HUMAN WEAKNESSES

I've spoken about some of the disadvantages of AI when compared to human intelligence. Artificial intelligence may, for example, be more brittle and less semantic than human intelligence. But it is not all loaded in our favour. There are many ways that AI will be superior.

There are lots of problems on which AI has already exceeded human intelligence. The best chess and poker players on the planet are now computers. A computer-vision algorithm can read chest X-rays more quickly, more accurately and far more cheaply than a human doctor can. And the fastest solver of the Rubik's cube is a robot, not a person.[9]

Some of the advantages of AI over human intelligence are physical characteristics. A computer can be faster than a human. It works at electronic speed, much faster than the slow chemistry of the human brain. A computer can have more memory than a human brain. And a computer can call upon more energy than a human brain.

But some of the advantages of AI will be down not to its own strengths but to the weaknesses of human intelligence. Behavioural economics, for instance, is a catalogue of examples of human decision-making that are far from optimal. Economists consider *Homo economicus* a perfectly rational and optimal decision-maker. The reality is that humans are neither rational nor optimal.

One example is loss aversion. Consider tossing a fair coin: if the coin lands on heads, you win $1001, but on tails you lose $1000. This is a certain win for *Homo economicus*. Play the game ten times and *Homo economicus* expects to be $10 ahead. But most people will avoid tossing the coin even once, since our psychological response to losses is greater than it is to gains.

Another example is risk aversion. Consider tossing another fair coin, but this time if the coin lands on heads, you win $20, but on tails you lose $10. Can I pay you $9 to play the game in your place? If you play the game you'll win, on average, $10 more than you lose. But most people prefer certainty over risk, and so will take the $9 instead. This is not the choice that the rational *Homo economicus* makes.

Artificial intelligence may thus be a route to more rational and better decision-making. Indeed, this is why IBM's AI program Watson was able to beat humans at the quiz show *Jeopardy!*. It didn't win because it was better at answering the general knowledge questions. It wasn't. Watson won in large part because it was better than humans at betting on the answers.[10]

In a similar way, we should contemplate the idea that computers could behave more ethically than humans. There are many ways this might be possible. They are, for example, not afflicted by human weaknesses. As a result, they can act more altruistically, always choosing to sacrifice themselves to save a human life.

More fundamentally, we should ponder the possibility that computers not just *could* behave more ethically but *should* behave more ethically than humans. If we can hold them to higher standards, then aren't we morally obliged to do so? In the next chapter, we'll consider how we might come up with appropriate ethical rules to ensure that computers behave more ethically than humans.

ETHICAL RULES

THE LAST INVENTION

I.J. Good was a brilliant mathematician who worked with Alan Turing at Bletchley Park and later at Manchester University on some of the very first computers. He advised Stanley Kubrick on the making of the film *2001: A Space Odyssey*. He famously wrote to the Queen suggesting that he be made a peer of the realm, because then people would say, 'Good Lord, here comes Lord Good.'[1] Less humorously, Good made a simple but rather worrying prediction about humanity's last invention:

> The survival of man depends on the early construction of an ultraintelligent machine ... Let an ultraintelligent machine be defined as a machine that can far surpass all the intellectual activities of any man however clever. Since the design of machines is one of these intellectual activities, an ultraintelligent machine could design even better machines; there would then unquestionably be an 'intelligence explosion,' and the intelligence of man would be left far behind. Thus the first ultraintelligent machine is the last invention that man need ever make, provided that the

machine is docile enough to tell us how to keep it under control. It is curious that this point is made so seldom outside of science fiction. It is sometimes worthwhile to take science fiction seriously ...[2]

To deal with this control problem, Good proposed a simple ethical rule that might ensure a good outcome for humanity.[3] He suggested the following direction that any ultra-intelligent machine should observe: 'Treat your inferiors as you would be treated by your superiors.'

This rule is not due to Good. It is, in fact, around 2000 years old. It comes from Seneca the Younger, a Stoic philosopher who was the emperor Nero's adviser in ancient Rome.[4] History perhaps tells us that Nero didn't listen to this part of Seneca's advice: his reign was a time of great tyranny, extravagance and debauchery.

Despite its simplicity and elegance, I fear that Good's rule would not ensure an ultra-intelligent machine behaves well. For one, we don't want a robot to treat us exactly like they want to be treated. We certainly don't want to be connected to 240 volts at regular intervals, or powered down at night. But even metaphorically, this rule is problematic. We humans don't want to do all the dull, dirty, difficult and dangerous things that we will get robots and other AIs to do. And I repeat my earlier suggestion that we should, in fact, demand higher ethical standards for AI than for humans. Because we can. Shouldn't an ultra-intelligent robot be required to sacrifice itself for us? Shouldn't an ultra-intelligent computer be without the unconscious biases that even the most careful of us have?

FICTIONAL RULES

Perhaps the best-known ethical rules for AI have come from science fiction. In 1942, Isaac Asimov proposed his famous laws of robotics.[5] These three laws require robots to protect themselves, unless this conflicts with an order from a human, and to follow any such order unless this might cause harm to a person.

Asimov's Three Laws of Robotics

First Law: A robot may not injure a human being or, through inaction, allow a human being to come to harm.

Second Law: A robot must obey the orders given it by human beings except where such orders would conflict with the First Law.

Third Law: A robot must protect its own existence as long as such protection does not conflict with the First or Second Laws.

Unfortunately, Asimov's stories illustrate that this well-crafted set of laws fails to cover all possible situations. For example, what happens if a robot must harm one human to save several others? What if both action and inaction will harm a human? What does a robot do if two humans give contradictory orders? Despite such concerns, Asimov argued that robots should follow his three laws, even before they become ultra-intelligent.

I have my answer ready whenever someone asks me if I think that my Three Laws of Robotics will actually be used to govern the behavior of robots, once they become versatile and flexible enough to be able to choose among different courses of behavior. My answer is, 'Yes, the Three Laws are the only way in which rational human beings can deal with robots – or with anything else.' But when I say that, I always remember (sadly) that human beings are not always rational.[6]

Notwithstanding Asimov's strong belief in his laws, I, like many of my colleagues, remain sceptical that they are sufficient to ensure robots behave ethically. Asimov himself conceded that humans are not rational, and that robots will have to cope with our irrational behaviours. Equally, his laws are imprecise and incomplete. It is a big challenge to provide precision and cover circumstances we might never imagine. In the course of developing its self-driving cars, for example, Google has experienced some bizarre and unexpected situations. A Google self-driving car once came across an elderly woman in a motorised wheelchair who was waving a broom around at a duck she was chasing down the street. Wisely, the car stopped and refused to go on.[7]

One feature of Asimov's laws that is often overlooked is that they are supposed to be hard-wired into a robot's positronic brain. There should be no way to circumvent them. Whatever ethical rules are embedded in our robots and other AI systems also need to be hard-wired. Machine learning is often a major component of AI systems. And in machine learning, the program is learned from the data and changes over time. It is not explicitly programmed by some human. We need, therefore, to be careful that the system doesn't learn to act unethically.

RESPONSIBLE ROBOTS

For the last 80 years, Asimov's laws have largely been ignored by those actually involved in building AI and robotics. They have remained science fiction rather than science fact. However, in the last decade it has become clear that storm clouds have been brewing. Many people, me included, have started to think seriously about the need to ensure that robots don't go rogue.

In 2009, Robin Murphy, a professor of robotics at Texas A&M University, and David Woods, a professor at Ohio State University working on improving the safety of systems in high-risk complex settings, proposed 'The Three Laws of Responsible Robotics'. Their goal was not to provide a definitive set of ethical rules but to stimulate discussion.

Their new rules didn't diverge much from Asimov's, and so haven't advanced the conversation much. However, they did make it clear that the responsibility rests on humans. Woods put it plainly: 'Our laws are a little more realistic [than Asimov's Three Laws], and therefore a little more boring.'[8]

The Three Laws of Responsible Robotics

1. A human may not deploy a robot without the human–robot work system meeting the highest legal and professional standards of safety and ethics.

2. A robot must respond to humans as appropriate for their roles.

3. A robot must be endowed with sufficient situated autonomy to protect its own existence as long as such protection provides

smooth transfer of control which does not conflict with the first and second Laws.[9]

In 2010, there was a more ambitious attempt across the Atlantic to advance the conversation on robot rules. The main UK government body for funding AI research, the Engineering and Physical Sciences Research Council (EPSRC), along with the Arts and Humanities Research Council (AHRC), brought together a small group of experts to consider rules for developing robotics both responsibly and for the maximum benefit of society. The group included specialists in technology, the arts, law and social sciences. The meeting resulted in the publication of five principles for robotics, which expanded on Asimov's three laws. The five rules were not intended as hard-and-fast laws but as a living document. The goal was to inform debate.

EPSRC/AHRC Five Principles of Robotics

Principle 1. Robots are multi-use tools. Robots should not be designed solely or primarily to kill or harm humans, except in the interests of national security.

Principle 2. Humans, not robots, are responsible agents. Robots should be designed; operated as far as is practicable to comply with existing laws & fundamental rights & freedoms, including privacy.

Principle 3. Robots are products. They should be designed using processes which assure their safety and security.

Principle 4. Robots are manufactured artefacts. They should not be designed in a deceptive way to exploit vulnerable users; instead their machine nature should be transparent.

Principle 5. The person with legal responsibility for a robot should be attributed.

The first three of these five principles reflect Asimov's laws. The first principle, like Asimov's First Law, aims to prevent robots from harming humans. Except it has a worrying get-out clause for national security. That is a disappointing inclusion. Couldn't national security be served by non-lethal robots?

The second principle, like Asimov's Second Law, concerns responsibility. And the third principle, like Asimov's Third Law, addresses safety and security. The other two principles introduce some new and important ideas. The fourth principle considers deception and transparency. Both have become important parts of many conversations about AI and ethics. And the fifth and final principle concerns legal responsibility and accountability.

Aside from the exemption that robots can kill in the interests of national security, it is hard to disagree with these five principles. But they leave open many questions. Who is responsible for a robot that has learned bad behaviours from a third party: the owner, the manufacturer, the third party, or some combination of the three? The fourth principle states that robots should not be deceptive to vulnerable users. Does this mean robots can be deceptive to users who aren't vulnerable? Should robots ever be made in human form, as this hides their machine nature?

The idea that robots can kill in the interests of national security is very problematic. Are there any limits on what robots can do in the interests of national security? Can they torture a confession out of a suspected terrorist? Or do human laws apply to robots? National security does not override the fundamental human right to life.

Others have continued to add detail to such robot rules. In 2016 the British Standards Institution published the first explicit national rules for robots: *BS 8611 Robots and Robotic Devices: Guide to the Ethical Design and Application of Robots and Robotic Systems*. This is a national standard providing detailed guidance to robot designers on assessing and mitigating the ethical risks associated with robots.

The guidelines list 20 different ethical hazards and risks in a range of domains, including societal, commercial and environmental. They address the safe design of robots, as well as ways to eliminate or reduce risks to acceptable levels. The risks identified include loss of trust, deception (again), invasion of privacy, as well as more wide-ranging concerns such as addiction and loss of employment. The *Guardian* newspaper summarised the 28-page standard with a beautifully short headline: 'Do No Harm. Don't Discriminate.'[10]

THE ACADEMY SPEAKS

In January 2017, I was invited to join over 100 leading researchers in AI, economics, law and philosophy, along with some well-known figures like Elon Musk, at a meeting to discuss AI and ethics at the Asilomar Conference Center. This sits on the beautiful shores of the Pacific Ocean in Monterey, California. The location was carefully chosen – not for its great natural beauty, but for its history and what it signalled.

In 1975, a similar conference had brought researchers to Asilomar to discuss the risks of the emerging science of recombinant DNA. At this meeting, the assembled scientists agreed on an influential set of guidelines to conduct experiments safely and to limit the possible escape of a biohazard into the environment. The 1975 gathering marked the beginning of a more open era in science policy. It set a powerful precedent for scientists to identify *in advance* the risks introduced by a new technology, and to put safeguards in place to protect the public.

The goal of the 2017 meeting was therefore clear to the participants when the invitation arrived in our inboxes. At the end of the meeting, we voted and agreed upon the 23 Asilomar AI Principles.[11] You might have spotted an inflationary trend here. We went from Good's single rule to Asimov's three laws, to the five principles of responsible robotics, and finally to the 23 principles agreed upon at Asilomar.

The Asilomar principles divide into three broad areas: research issues, ethics and values, and longer-term issues. They cover a wide range of ethical concerns, from transparency through to value alignment to lethal autonomous weapons and existential risk.

Like at the 1975 Asilomar meeting on recombinant DNA, the precautionary principle drove many of our discussions. The precautionary principle is a powerful ethical and legal approach to dealing with changes that have the potential to cause harm when scientific knowledge on the matter is lacking.[12] The precautionary principle, as its name suggests, emphasises caution in pushing forward activities that may prove very harmful, especially when those might be irreversible.

The precautionary principle breaks down into four interrelated parts. First is the idea that scientific uncertainty requires the regulation of activities that pose a risk of significant damage. Second is the idea that any such regulatory controls should incorporate a margin of safety. Third, when the harm can be great, the best available technology should be used to prevent any poor outcomes. And fourth, activities that present an uncertain potential for very significant, perhaps irreversible injury should be prohibited.

When you are considering the long-term risks of AI, such as super-intelligence, the precautionary principle is a reasonable approach. However, even for more immediate risks, the precautionary principle informed much of our thinking at Asilomar. For instance, given our uncertainty about the short-term societal impacts of AI, the second Asilomar principle recommends funding social scientists and others to consider such impacts.

The Asilomar AI principles have not had the impact of the earlier Asilomar principles on recombinant DNA. It didn't help that the meeting had a bias towards AI researchers from the United States and Europe, and not from further afield. Nor did it help that the meeting had a bias towards academic AI researchers, and didn't include many people from the Big Tech companies that are actually deploying AI.

Nevertheless, it's hard to imagine a better outcome from a three-day meeting like this. AI introduces a broad range of ethical issues, from near-term concerns like privacy to long-term challenges such as super-intelligence, from tricky economic concerns such as technological unemployment to difficult military matters such as lethal autonomous weapons, from important environmental problems such as sustainability to thorny welfare issues like inequality. There was little chance that we could solve all of these in three days of discussion.

The 23 Asilomar principles may look as if they were written by a committee. But that's to be expected. Because they were written by a committee. Everything we could think of was thrown into the pot. As I recall, almost nothing that was introduced was thrown out. It is also not hard to pick holes in many of the principles. Why, for example, do we only worry about value alignment in highly autonomous systems? Personally, I worry about value alignment in the simple algorithms used by Facebook and Twitter. And how can we possibly achieve many of the desired outcomes of the 23 principles, such as shared prosperity or human dignity?

Despite these criticisms, the Asilomar AI principles have had some positive effects. In particular, politicians have started to take more notice of ethical concerns around AI, especially in Europe.

EUROPE LEADS

Perhaps more than any other body, the European Union has been at the forefront of ensuring the responsible use of AI. In June 2018, the European Commission announced the High-Level Expert Group on AI. This is a group of 52 experts drawn from academia, civil society and industry, and appointed by the EU

Commission to support the implementation of the European strategy on artificial intelligence.

In April 2019, following extensive public consultation, the High-Level Expert Group on AI presented seven 'key ethical requirements' for deploying trustworthy artificial intelligence.[13]

1. *Human agency and oversight.* AI systems should empower human beings, allowing them to make informed decisions and fostering their fundamental rights. At the same time, proper oversight mechanisms need to be in place.

2. *Technical robustness and safety.* AI systems need to be resilient and secure. They need to be safe, with a fall-back plan in case something goes wrong, as well as being accurate, reliable and reproducible. This will ensure that unintentional harm is minimised and prevented.

3. *Privacy and data governance.* Besides ensuring full respect for privacy and data protection, adequate data governance mechanisms must also be in place, taking into account the quality and integrity of the data, and ensuring legitimate access to data.

4. *Transparency.* The data, system and AI business models should be transparent. Traceability mechanisms can help achieve this. Moreover, AI systems and their decisions should be explained in a manner adapted to the stakeholder concerned. Humans need to be aware that they are interacting with an AI system, and must be informed of the system's capabilities and limitations.

5. *Diversity, non-discrimination and fairness.* Unfair bias must be avoided, as it has multiple negative implications, from the marginalisation of vulnerable groups to the exacerbation of prejudice and discrimination. To promote diversity, AI systems should be accessible to all, regardless of any disability, and involve relevant stakeholders throughout their entire life cycle.

6. *Societal and environmental well-being.* AI systems should benefit all human beings, including future generations. They must therefore be sustainable and environmentally friendly. Moreover, they should take into account the environment, including other living beings, and their social and societal impact should be carefully considered.

7. *Accountability.* Mechanisms should be put in place to ensure responsibility and accountability for AI systems and their outcomes. Auditability, which enables the assessment of algorithms, data and design processes, plays a key role here, especially in critical applications. Moreover, adequate and accessible redress should be ensured.

The European Union has a strong and respected history in regulating the digital space. In 2016, the European Parliament and Council of the European Union passed the *General Data Protection Regulation* (GDPR). This law came into force in May 2018. Despite initial concerns, it has proven effective in restoring some personal privacy to consumers. California passed a similar data protection law that took effect on 1 January 2020. In this case, as Oscar Wilde observed, imitation is the sincerest form of flattery.[14]

The European Commission, the executive arm of the European Union, has also imposed record fines on the Big Tech companies to promote better behaviours. In total, it has handed out over $13 billion in penalties over the last five years. The biggest fine, of more than $5 billion, was imposed on Google for anti-competitive behaviour in its distribution of the Android operating system. To put that $5 billion in perspective, this is 4 per cent of Google's annual revenue.[15]

The ethical guidelines that the EU came up with are very good. As you would expect if you bring together 52 of Europe's best experts in technology, law and other areas, and give them a year to deliberate on these matters. The guidelines leave many issues at a high level but that is probably of necessity. The challenge, then, is making the principles actionable.

How, for example, do you formulate hard and soft law so that AI systems are indeed transparent? To ensure accountability, do we need to mandate that autonomous systems such as self-driving cars have black boxes[16] that can record data leading up to any accident? And what precisely does it mean to design AI systems to avoid bias? How do we incentivise and regulate companies to build AI technologies that are inclusive? And how do we ensure AI systems take into account their societal impact?

THE ETHICS BANDWAGON

Many countries have jumped on the AI and ethics bandwagon, and put out their own national guidelines: Australia, the United

Kingdom, France, Germany, India, Japan, Singapore and Canada, to name some of the most prominent. International bodies like the G20, the United Nations High Commission for Human Rights and the World Economic Forum have also produced their own ethical frameworks. And 42 countries have adopted the Organisation for Economic Co-operation and Devlopment's (OECD) five value-based AI principles.

But it hasn't stopped there. Non-governmental organisations like Algorithm Watch, AI Now, AI4People, IEEE and the Institute for the Future have proposed yet more ethical guidelines. And a wide range of companies, including Google, Microsoft, SAP, IBM, Accenture and PricewaterhouseCoopers, have proposed their own ethical frameworks. Google has even said it will start selling Ethics as a Service to other companies.

Let me quote from one such set of ethical principles:

> AI development should promote fairness and justice, protect the rights and interests of stakeholders, and promote equality of opportunity. AI development should respect and protect personal privacy and fully protect the individual's right to know and right to choose. AI systems should continuously improve transparency, explainability, reliability, and controllability, and gradually achieve auditability, supervisability, traceability, and trustworthiness.

This all looks good – until you realise that these fine-sounding ethical principles were proposed by China's Ministry of Science and Technology.[17] It's unclear how the facial-recognition software being used by Chinese authorities to monitor and suppress Uighur people in Xinjiang Province is protecting human rights. Or how the voice-recognition software sold by China's leading AI unicorn, iFlytek, to the Chinese police to enable wide-scale monitoring of the population is respecting and protecting personal privacy.

It's hard not to conclude that words can be cheap in this space. Do we really need more ethical frameworks for the use of artificial intelligence? Are there actually any substantial differences between the ethical principles put forward in the different frameworks? And how do we go beyond such smoke and mirrors and give the public real protections?

HUMAN, NOT ROBOT RIGHTS

One possible approach is to base ethics for AI on human rights. The argument goes that many of the ethical issues around AI concern human rights, such as the right to equality, the right to privacy and the right to work. And if we are looking for international consensus or existing legislation that might apply, then human rights is one place where such consensus has already been found and rules enacted.

Now, AI definitely needs to protect human rights. We need more, not less, respect for the sort of fundamental concerns that arise from considering the impact of AI on human rights. But human rights are a lower bound on what we should be seeking. And I'm very doubtful that we should be looking to regulate AI at the international level, like we have regulated many human rights.

Ethics requires trade-offs. There are, for example, basic tensions between the rights of the individual and the rights of a group. Your freedom of speech may infringe on my right to privacy. How we resolve these tensions depends on the country in which we live. For instance, compared to a country like China, the United States might put the rights of the individual higher than those of the wider society.

This is precisely why there are so many different AI ethical frameworks. Each framework places a different emphasis on particular ethical values. How do we prioritise between fairness, transparency, explainability, privacy and robustness? There is not one solution – and certainly not one that will be agreed at the international level.

History provides a good analogy here. AI is often compared to electricity. Like electricity, AI will be a pervasive technology that is in all our homes, offices and factories. Like electricity, AI will be in almost every device. Indeed, it will *be* the operating systems of those devices, providing the smartness in our smart speakers, our smart fridges and our smart cars. If we go back a century or so, the electricity revolution changed our planet, much like the AI revolution is starting to change it today. But we did not regulate electricity internationally. It would have been relatively easy to agree on voltages and frequencies, even the number and shape of the pins on a plug. But for various reasons we didn't do that.

AI will be much more complex and difficult to regulate than electricity. It's hard to imagine we'll reach meaningful global consensus on the many thorny issues concerning AI. What limits, for example, should we put on the (mis)use of facial-recognition software? What precisely does it mean for some decision-making software to be 'fair'? How do we ensure self-driving cars are safe enough and reliable enough to be on public roads?

THIS ISN'T THE FIRST TIME

If a focus on human rights isn't the best approach, what is? Perhaps history can provide some more clues. Artificial intelligence isn't the first technology to touch our lives, and every other technology we've invented has introduced ethical challenges. Take, for example, the invention of the printing press by Johannes Gutenberg in the fifteenth century. This was undoubtably one of the most transformative technologies of the last 500 years. And yet, as Mark Twain wrote in 1900, not everything it brought has been for the betterment of humankind:

> All the world acknowledges that the invention of Gutenberg is the greatest event that secular history has recorded.
>
> Gutenberg's achievement created a new and wonderful earth, but at the same time also a new hell. During the past 500 years Gutenberg's invention has supplied both earth and hell with new occurrences, new wonders and new phases.
>
> It found truth astir on earth and gave it wings; but untruth also was abroad, and it was supplied with a double pair of wings.
>
> Science was found lurking in corners, much prosecuted; Gutenberg's invention gave it freedom on land and sea and brought it within reach of every mortal.

Arts and industries, badly handicapped, received new life. Religion, which, during the Middle Ages, assumed tyrannical sway, was transformed into a friend and benefactor of mankind.

On the other hand, war, which was conducted on a comparatively small scale, became almost universal through this agency. Gutenberg's invention, while having given to some national freedom, brought slavery to others.

It became the founder and protector of human liberty, and yet it made despotism possible where formerly it was impossible.

What the world is to-day, good and bad, it owes to Gutenberg. Everything can be traced to this source, but we are bound to bring him homage, for what he said in dreams to the angered angel has been literally fulfilled, for the bad that his colossal invention has brought about is overshadowed a thousand times by the good with which mankind has been favored.[18]

Can we perhaps learn from the history of past inventions such as the printing press? In fact, despite its immense impact, the printing press is not a good example to study. When it was invented, there appears to have been little consideration of the upheavals that the printing press would help bring about.

Books have changed the world in profound ways. Political books like Karl Marx and Friedrich Engels' *The Communist Manifesto* and Thomas Paine's *The Rights of Man* challenged how we run our society. Economic texts like Adam Smith's *The Wealth of Nations* and John Maynard Keynes' *The General Theory of*

Employment, Interest and Money changed how we run our economies. And scientific books like Darwin's *On the Origin of Species* and Newton's *Principia Mathematica* transformed our understanding of the world around us. It's hard to imagine anyone predicting this in the fifteenth century, when the printing press was first used to distribute Bibles more cheaply.

MEDICAL LESSONS

Let me turn instead to a different field where we have worried greatly about the impact of new technologies on people's lives – medicine – as it may serve as a better model for AI. It's not surprising that ethics has been a major concern in medicine, as doctors are often dealing with life-or-death situations. As a result, medicine has some very well-developed ethical principles to guide how technology touches our lives.

Indeed, I will argue that, if we put to one side the thorny issue of machine autonomy, medicine provides an otherwise adequate set of ethical principles to guide the development of artificial intelligence.[19] There are four core ethical principles that have been developed over the last two millennia to steer medical practice.

The first two principles commonly considered in medical ethics are *beneficence* and *non-maleficence*. These are closely related. Beneficence means 'do good', while non-maleficence means 'do no harm'. Beneficence involves balancing the benefits of a treatment against the risks and costs it will bring. A medical intervention that has a net benefit is considered ethical. Non-maleficence, on the other hand, means avoiding harm. Of course, it may not be totally avoidable, but any potential harms should be proportionate to the potential benefits.

Many of the AI principles put forward in the European guidelines, the Asilomar principles or many of the other proposed frameworks follow, in fact, from ensuring beneficence and non-maleficence. For instance, robustness is needed to ensure AI doesn't do unnecessary harms. Invasion of privacy is a common harm that AI may enable. And insisting that AI systems should benefit all humans, as well as take care of the environment, follow from beneficence.

The third principle commonly considered in medical ethics is *autonomy*. It requires practitioners to respect the right of people to make informed decisions about their own medical care. Consent from a patient is essential before any medical treatment. And patients need to understand all the risks and benefits, and to be free from coercion as they make decisions.

Again, many of the AI principles put forward in the European guidelines and elsewhere follow from valuing the autonomy of humans as they interact with an AI system. Human agency and oversight, for example, follow from a respect for human autonomy. Other principles like transparency enable autonomy to be achieved. And respect for human autonomy explains why deceptive AI is to be avoided.

The fourth and final principle commonly considered in medical ethics is a somewhat fuzzy notion of *justice*. This obliges us to distribute fairly benefits, risks, costs and resources. In particular, the principle of justice requires both the burdens and the benefits of new medical treatments to be distributed equally across all groups in society.

As before, many of the AI principles suggested in the European guidelines and elsewhere follow from seeking justice. AI systems should be fair and not discriminate. AI systems should also be transparent and provide explanations so that justice can be seen to have been achieved.

Of course, AI is not medicine. The four ethical principles commonly used in medicine are a very good start, but by far not the end of what we need. Compared to medicine, artificial intelligence does not have the common aims and fiduciary duties found in medicine. AI also lacks the long and rich professional history and norms found in medicine, which ensure that these ethical standards are upheld. In addition, AI needs the robust legal and professional structures found in medicine to ensure accountability.

POWERFUL CONCERNS

As the discussion around AI and ethics has become more sophisticated, some of the conversation has shifted to considerations of power. Who benefits from the system? Who might be harmed by it? Can people opt out? Does the system discriminate, perhaps increasing the systemic inequalities already afflicting our society? Is the system making the world a better place?

These are all important questions to ask. But power does not trump ethics. And power is not always a bad thing. Power can be benevolent. From Gandhi to Mandela, there are leaders who have exercised power without seeking to harm others. Equally, those lacking power can still behave poorly. And harms may occur despite the absence of any power dynamics.

Focusing on power rather than ethical concerns brings other risks. Issues that weren't struggles for power may become such. Friendly voices within those power structures may become alienated, even accused of the very harms that they have been trying to avoid from their privileged positions. Arguments about power often turn the debate into a competition between winners and losers, and not into a discussion about how everyone can win.

There are many other ethical traps to avoid.[20] Being fair (or transparent, or any other value) doesn't mean you're being ethical. Ethics cannot be reduced to a simple checklist. There is no universal set of ethical values with which we need to align our AI systems. Similarly, it is not a simple dichotomy between systems that are ethical and systems that are unethical. And we need to be sensitive to the fact that ethical trade-offs will change as the context changes. These are only some of the many pitfalls to avoid.

The eminent philosopher Daniel Dennett is pessimistic about the ability of AI systems to act in ethically sound ways:

> AI in its current manifestations is parasitic on human intelligence. It quite indiscriminately gorges on whatever has been produced by human creators and extracts the patterns to be found there – including some of our most pernicious habits. These machines do not (yet) have the goals or strategies or capacities for self-criticism and innovation to permit them to

transcend their databases by reflectively thinking about their own thinking and their own goals.[21]

It is clear that we have a long way to go. But this is to be expected. These are complex and difficult ethical questions that we have been grappling with for thousands of years. The mistake is to think we can come up with quick and easy answers. One area in which AI has thrown up some of the most pressing ethical questions is fairness. In the next chapter, I consider how algorithms can help or hurt the fairness of our decision-making.

FAIRNESS

There are plentiful examples of problems involving the fairness of AI-enabled decision-making. To compound matters, we don't yet have a precise playbook for fixing them. In part, this is because it is still early days in dealing with some of these issues. But it is also because there probably aren't going to be simple fixes for many of these problems.

Fairness goes to the heart of what it means to live in a just and equitable society. This is something we've struggled with ever since we started living alongside each other, and the answers we have continue to evolve. AI puts some of these fairness issues on steroids. However, despite there not being good answers to many of these questions, there are a dozen valuable lessons that I will pull out.

Before we look into these challenges, I want to mention the many possible upsides to automating decision-making that could help make the world a fairer place. First, handing over decisions to computers could lead to greater consistency. Humans can be capricious and random in coming to decisions. Computer programs, on the other hand, can be frustratingly consistent. We are often most aware of this when they're consistently doing the wrong thing.

Second, automated decision-making has the potential to be more transparent than human decision-making. Humans are far from transparent in how they make decisions, and it's not certain we'll ever truly understand and be able to record how we do so. Even though many automated systems are not easily understood today, there are no fundamental reasons why we can't make them more transparent in the future.

Third, human decision-making is full of unconscious biases. We can work hard at eliminating these, but even the best of us struggles to do so. All of us subconsciously make decisions based on gender, race and other attributes, even though we know we shouldn't and try hard not to do so. When we automate decisions, we can simply not include those attributes in the data given to the machine. Eliminating bias isn't as simple as this, but this may at least be a first step to fairer decisions.

Fourth, and perhaps most crucially, automated decision-making can be much more data-driven and evidence-based. There are many situations where humans make decisions based on intuition. But in many of these settings, we are now collecting and analysing data. We can therefore determine for the first time whether our decisions are fair. And when they are not, we can consider adjusting the decision-making to improve the fairness.

MUTANT ALGORITHMS

In August 2020, we saw what were perhaps the first (but I suspect not the last) protest marches over an algorithm. Due to the COVID-19 pandemic, students at UK schools weren't able to sit A-level or

GCSE exams. Instead, an algorithm was used to allocate grades based on the predictions of teachers.

According to the official regulator Ofqual, the algorithm adjusted around four in ten marks down by one grade or more from the grades predicted by teachers. Students took to the streets of London to protest. The government quickly capitulated and reverted to the grades predicted by teachers. The prime minister, Boris Johnson, blamed the fiasco on 'a mutant algorithm'. But there was nothing mutant about the algorithm. As far as we can tell, it did exactly what it was meant to do. The problem was that Ofqual hadn't thought carefully enough about what the public would find *fair*.

We shall never know how accurate the algorithm was, since the students didn't sit their final exams. But it is worth pointing out that teachers aren't that good at accurately estimating the performance of their students. In Scotland, teacher estimates are collected every year, and only 45 per cent of students in a normal year achieve their estimated grade. We shouldn't expect teachers to have done any better in estimating grades in an exceptional year like 2020. It is also worth pointing out that human markers often don't agree with each other. Outside of maths and the sciences, 30 per cent of A-level markers disagree on what grade to give a paper.

So let's be generous and not too critical of the accuracy of Ofqual's algorithm. After all, the algorithm was designed to give an overall distribution of grades that looked similar to previous years, with similar proportions of grades for each subject. In fact, mirroring recent years, it even allowed a slight increase in A and A* grades. Ofqual even went so far as to check that the proportion of grades handed out to different subpopulations (by gender, ethnicity and income, for example) matched that in recent years.

However, what soon became apparent was that even if Ofqual's algorithm was as accurate as human markers, it was nevertheless biased in favour of certain groups and against others. In particular, it was biased in favour of pupils at schools that had done well in previous years, and in subjects where class sizes were small. Or, to put it another way, it was biased against students in poor public schools and in favour of students at rich private schools.

How did this happen? The algorithm tried to ensure that the range and distribution of grades achieved by a group of students in a class was similar to students in the same class over the previous three years. The exception to this was small classes – those with less than 15 students in most cases – where the teacher-predicted grades were used because of the lack of a data algorithm. The net result was that the algorithm was fair *on average*, providing a similar distribution of grades as in past years. But the unfairness was highly biased against students in public schools, especially in more deprived areas, and highly biased in favour of students in private schools. Small teaching groups, and less popular A-levels such as Law, Ancient Greek and Music, are more common at private schools, which insulated those students from being marked down. And, historically, A-level results in selective and private schools have been higher, bequeathing a higher range of grades to the 2020 students.

I suspect that even if Ofqual had avoided these problems of bias, they would have run into trouble. Some students were going to win, some were going to lose. The winners won't shout out, but you can be sure the losers will complain loudly. Ofqual therefore had an impossible task. The government should have been generous, accepted that Ofqual was going to have to give out some higher grades, and funded more university places to compensate.

If we apply the sort of ethical principles used to inform decision-making in other areas like medicine, we can see that the marking algorithm failed the principle of justice. The burdens and benefits were not distributed equally across all groups in society. Students from poor state schools were more likely to have their grades marked down than students from rich private schools. There is no justice in this.

Actually, the marking algorithm exposed two more fundamental problems with the examination system. First, it highlighted how important it is to retain human agency, especially in high-stakes decisions. In the case of exams, people felt they had agency as they could hope to ace the exam. But having an algorithm give students a predicted grade without an exam, however accurate the prediction might have been, denied them this agency.

Second, the marking algorithm exposed and magnified a fundamental problem with the public examination system that had existed even when humans were doing the marking. Ranking students nationwide on a simple scale, when this ranking would decide life-changing events like university places, is inherently questionable. Should your life options be decided by how well you perform in an exam on one particular afternoon? That seems no better than the whims of an algorithm. The truth is that algorithms cannot fix broken systems. They inherit the flaws of the systems into which they're placed.

PREDICTIVE POLICING

Another problematic area where AI algorithms have been gaining traction is in predictive policing. This sounds like the film *Minority Report* but is actually far simpler. We cannot predict when someone is going to commit a crime. Humans are not that predictable. But we can predict where, *on average*, crime will take place.

We have lots of historical data about crime: police incident reports, sentencing records, insurance claims and so on. And in most districts there are insufficient police resources to patrol all the neighbourhoods we would like. So why not use machine learning to predict where and when crimes are most likely to take place, and focus police resources on those places and times?

In late 2011, *Time* magazine identified predictive policing (along with the Mars Rover and Siri) as one of the year's 50 best inventions.[1] It is now used by police departments in a number of US states, including California, Washington, South Carolina, Alabama, Arizona, Tennessee, New York and Illinois. In Australia, the NSW police force has an even more sinister and secret algorithm, which predicts not when and where crime might take place, but who is likely to commit a crime. These individuals are then given extra scrutiny and monitoring by local police. A report analysing police data found that people under the age of 25, along with Indigenous citizens, were disproportionately targeted, and that the algorithm makes decisions based on 'discriminatory assumptions'.[2]

There are several fundamental problems here. First, we don't have ground truth. We want to predict where and when crime will take place, and who is going to be responsible, but we simply can't know this. We only have historical data on where crime was reported. And there is a lot of crime that took place that we don't know about.

The data we have reflects the biases of the system in which it was collected. Perhaps the police disproportionately patrolled poorer neighbourhoods. The greater prevalence of crime reported in such neighbourhoods might therefore be simply a consequence of this greater number of patrols. Or it might be the result of racism within the police force, which meant that more Black people in such neighbourhoods were stopped and searched.

Predicting future crime on such historical data will then only perpetuate past biases. Let me adapt a famous quotation from writer and philosopher George Santayana: 'Those who use AI to learn from history are doomed to repeat it.' In fact, it's worse than repetition. We may construct feedback loops in which we magnify the biases of the past. We may send more patrols to poorer and predominately Black neighbourhoods. These patrols identify more crime. And an unfortunate feedback loop is set up as the system learns to send even more patrols to these neighbourhoods.

SENTENCING

AI algorithms have been gaining traction in another part of the judicial system, helping judges decide who might or might not reoffend. There are dozens of risk-assessment algorithms now in use around the world. One of the most troubling is the COMPAS tool.[3] This was developed by Northpointe and, despite significant concerns about its fairness, remains widely used across the United States, helping judges decide the risk of a person reoffending.

In 2016, the investigative news organisation *ProPublica* published a damning study on the accuracy of the COMPAS tool.[4] They found that 'blacks are almost twice as likely as whites to be labeled a higher risk but not actually re-offend', whereas the tool 'makes the opposite mistake among whites: They are much more likely than blacks to be labeled lower-risk but go on to commit other crimes'.

Even ignoring these racial injustices, *ProPublica* found that the tool wasn't actually very good at predicting those likely to reoffend. Only one in five people it predicted would commit a violent crime actually went on to do so. When considering all possible crimes, not just violent crimes, it wasn't much better than a coin toss. Only 61 per cent of those predicted by the tool to reoffend were arrested within two years.

Northpointe pushed back against *ProPublica*'s claims, disputing both their analysis and their conclusions. Before I go into the details, let me discuss the broader moral issue: should we even use such a tool in the first place? Suppose we could build a tool that could predict more accurately than human judges those likely to reoffend – wouldn't we be morally obliged to use it?

Literature provides, I believe, a good answer to this question. I, for one, do not want to wake up in such a world, where machines get to decide our liberty. There are plenty of stories that have painted a dark picture of such a world. There is a good reason that we hand over some of the most consequential decisions in our society to human judges. We should not change this without careful consideration. Human judges can be held accountable for their decisions. Machines cannot.

Many proponents of tools like COMPAS will argue that they are only used to advise judges, and that humans are still ultimately making the decisions. However, there is plentiful psychological evidence that humans are strongly biased by suggestions from automated tools, and will overlook contradictory information even when it is correct. This phenomenon is called 'automation bias'. We tend to trust what a computer tells us even when it conflicts with other information we might have.

Let me give a classic example. In June 1995, the second officer of the *Royal Majesty* cruise liner put too much trust in the computerised course plotter, ignoring information that conflicted with its positioning of the ship – such as a lookout who identified 'blue and white water dead ahead'. Unfortunately for the second officer, the antenna to the GPS had come loose, so the plotter was using dead reckoning and not an accurate satellite fix. Due to the strong tides and winds, the *Royal Majesty* was 17 miles off course and in the middle of the notorious Rose and Crown shoal off Nantucket Island. The ship spent a day hard aground before tugs towed her off.

There is a simple reason to expect automation bias with a sentencing tool like COMPAS. Judges have a natural tendency to err on the side of caution and follow the advice of any tool. Especially in the United States, where many judges are elected, who wants to take the risk of releasing a felon when an automated tool has predicted they will reoffend? If it becomes public that the judge overlooked the advice of the tool before the felon reoffends, it will hurt their chances of being re-elected.

Putting aside, then, the strong moral arguments against handing over sentencing decisions, explicitly or implicitly, to a computer, there are strong sociological and technical arguments for definitely not using a sentencing tool like COMPAS which is so poorly designed and makes such poor decisions. Let me comment first on the design.

The inputs to COMPAS are the answers to 137 questions. The designers of COMPAS clearly didn't want the software to be racist, so race is not one of the inputs. But ZIP code is. And in many parts of the United States, ZIP code is a good proxy for race. Any decent machine-learning tool will quickly discover the correlations between race and ZIP code.

Many of the other inputs to COMPAS are also troubling. Have a look at some of the questions that defendants must answer:

31. Which of the following best describes who principally raised you?
    O Both Natural Parents
    O Natural Mother Only
    O Natural Father Only
    O Relative(s)

○ Adoptive Parent(s)

○ Foster Parent(s)

○ Other arrangement

32. If you lived with both parents and they later separated, how old were you at the time?

○ Less than 5

○ 5 to 10

○ 11 to 14

○ 15 or older

○ Does Not Apply

35. Were your brothers or sisters ever arrested, that you know of?

○ No

○ Yes

55. How often have you moved in the last 12 months?

○ Never

○ 1

○ 2

○ 3

○ 4

○ 5+

95. How often did you feel bored?

○ Never

○ Several times/mo[nth]

○ Several times/w[ee]k

○ Daily

97. How much do you agree or disagree with the following –

You feel unhappy at times?

O Strongly Disagree

O Disagree

O Not Sure

O Agree

O Strongly Agree

Do we really want to decide sentencing based on whether someone is bored or unhappy? Or whether other members of their family have been arrested? Or on matters that they likely had little control over, like being orphaned or having to move house when a landlord sells a property? At this point, you have to ask: what were the developers of COMPAS thinking?

This highlights a common mistake in machine learning where we confuse *correlation* with *causation*. It may well be that people who commit crimes tend to come more often from broken homes. Crime and a difficult childhood might be correlated in this way. But it is wrong to suppose causation – that is, that a difficult childhood *causes* crime to take place.[5] And it is even more wrong to punish those who had the misfortune of having had a difficult childhood.

We should be careful not to build AI systems which are based on such confusion. What a great injustice to someone who has pulled themselves up despite the many challenges of a tough childhood to then face institutional and systemic persecution from an automated tool?

PREDICTION ERRORS

A fundamental problem with COMPAS is that its predictions are not very accurate. Perhaps most damning is that there are two much simpler and less controversial methods that make predictions just as accurate as those of COMPAS, the first of which is almost certainly less biased.[6]

The first method that does just as well as the COMPAS tool is a simple linear classifier.[7] It uses just two of the 137 features that COMPAS uses: age and number of historical priors. We don't need the other 135 features, many of them troubling, that COMPAS uses as input to reach its accuracy.

The second method that is just as precise and fair as COMPAS is to ask random people who have no expertise in criminal justice to make a prediction on Amazon's Mechanical Turk. You tell them a few facts about the defendant, such as sex, age and previous criminal history, but not their race. And then you pay them $1 to predict whether the person will reoffend within two years. The median accuracy for these predictions was 64 per cent, almost identical to the 65 per cent accuracy of COMPAS.

Let's leave aside the inaccuracy of the predictions made by COMPAS and focus instead on the racial bias. We are trying to predict the future, and the future is inherently unpredictable. Perhaps you can't be very accurate. But that's no excuse to be racially biased. The problem is that the errors made by COMPAS are biased against Black people and in favour of white people. In particular, COMPAS is more likely to recommend that Black people who won't reoffend be locked up. At the same time, COMPAS is more likely to recommend that white people who will reoffend be released into the community.

We should be outraged by this racial injustice. Black people are being unfairly incarcerated, and white people who should be locked up are being set free. How, then, could Northpointe rigorously defend the fairness of the COMPAS tool despite *ProPublica*'s complaints? It turns out there are several different mathematical definitions of fairness. Northpointe chose to focus on a different definition than *ProPublica*, and if you rely on this second definition, the COMPAS tool does okay.

In fact, there are now at least 21 different mathematical definitions of fairness in use by the machine-learning community.[8] And you can prove that many of these definitions of fairness are mutually incompatible.[9] That is, except in trivial settings (such as when the two groups are completely indistinguishable), no prediction tool can satisfy more than one definition of fairness at a time.

To understand how there can be so many different ways to define fairness, we need to view the predictions of a tool like COMPAS through the lens of a 'confusion matrix'. This is a summary of the number of correct and incorrect predictions made by the tool. We reduce the tool's performance to four numbers: the *true positives* (the number of people predicted to reoffend who actually do reoffend), the *true negatives* (the number of people predicted not to reoffend who don't reoffend), the *false positives* (the number of people predicted to reoffend who don't) and the *false negatives* (the number of people predicted not to reoffend who do).[10]

Confusion matrix	Did not reoffend	Did reoffend
Predicted not to reoffend	#True negative	#False negative
Predicted to reoffend	#False positive	#True positive

ProPublica obtained the COMPAS predictions for 2013 and 2014 from the Broward County Sheriff's Office in Florida. They chose Broward County because it is a large jurisdiction that used the COMPAS tool for pre-trial release decisions, and because Florida has strong open-records laws. *ProPublica* summarised the predictions made by the COMPAS tool with two confusion matrices: one for Black defendants and the other for white.

Black defendants	Did not reoffend	Did reoffend
Predicted not to reoffend	990	532
Predicted to reoffend	805	1369

White defendants	Did not reoffend	Did reoffend
Predicted not to reoffend	1139	461
Predicted to reoffend	349	505

ProPublica looked at a fairness measure called the 'false positive rate'. This is the percentage of people who did not reoffend and who were wrongly predicted to reoffend. In symbols, this is a simple fraction[11] derived from the first column of the confusion matrix:

$$\frac{\text{FALSE POSITIVE}}{\text{FALSE POSITIVE} + \text{TRUE NEGATIVE}}$$

These are the people paying the price of the tool's inaccuracy, the people likely to be incarcerated who could be safely released. For white defendants, the false positive rate was 349/(349+1139) or 23 per cent. However, for Black defendants, the false positive rate was 805/(805+990) or 45 per cent, almost twice as large.

Northpointe considered a difference fairness measure, called 'precision'. This is the percentage predicted to reoffend who actually did reoffend. In symbols, this is another simple fraction derived from the second row of the confusion matrix:

$$\frac{\text{TRUE POSITIVE}}{\text{TRUE POSITIVE} + \text{FALSE POSITIVE}}$$

This is the percentage of people likely to be incarcerated who should be. For white defendants, the precision was 505/(505+349) or 59 per cent. On the other hand, for Black defendants, the precision was 1369/(1369+805) or 63 per cent. The COMPAS tool thus offered a very similar precision for Black or white defendants. But while precision was similar for Black and white people, the errors had a disproportionately negative impact on Black people.

The false positive rate considers the two entries in the first column of the confusion matrix. Precision, on the other hand, looks at the two entries in the second row. We could also consider other parts of the confusion matrix. Take, for instance, the entries in the second column. 'Recall' is the percentage of people who did reoffend who were correctly predicted to reoffend. In symbols, this is a simple fraction derived from the second column of the confusion matrix:

$$\frac{TRUE\ POSITIVE}{TRUE\ POSITIVE\ +\ FALSE\ NEGATIVE}$$

For white defendants, the recall was 505/(505+461) or 52 per cent. On the other hand, for Black defendants, the recall was 1369/ (1369+532) or 72 per cent. The COMPAS tool therefore did a much worse job of correctly identifying white people who did reoffend. In fact, it was little better than a coin toss on white defendants.

Ultimately, no one of these different fairness measures is better than another. They each represent different trade-offs. Do we value individual liberty, and seek not to lock up people wrongly? Or do we value society more, and seek not to have people released into society who might reoffend, but at the risk of wrongly locking up too many people? These are difficult questions, and different societies will make different choices.

THE PARTNERSHIP

The Partnership on AI is a non-profit advocacy organisation that was established in late 2016, led by a group of AI researchers representing six of the world's largest technology companies: Apple, Amazon, DeepMind and Google, Facebook, IBM and Microsoft.

The Partnership has since expanded to include over 100 partners, including organisations from academia and civil society such as AI Now, the Berkman Klein Center, Data & Society, the Alan Turing Institute, the BBC, UNICEF and the *New York Times*.

The mission of the Partnership on AI is to shape best practices, research and public dialogue about AI's benefits for people and society. It has unfortunately achieved very little to date. Indeed, a cynic like me might observe that it's part of the 'ethics washing' efforts of the Big Tech companies. There is one exception, though, which is a very good study undertaken by the Partnership on AI on the use of risk-assessment tools in the US criminal justice system.[12]

The study puts forward ten requirements that risk-assessment tools ought to satisfy. The requirements divide into three categories: (i) technical challenges related to accuracy, validity and bias; (ii) interface issues that reflect the ways in which people in the criminal justice system understand and use these tools; and (iii) governance, transparency and accountability issues arising from the fact that these tools may automate decisions.

The ten requirements are a roadmap for how to proceed, but are not precise standards in themselves. We still need to decide complex ethical trade-offs (such as that between individual liberty and the safety of society, as we saw earlier). Nevertheless, it is worth going through the ten requirements, as they highlight the many technical and other challenges of using AI tools in this space. They illustrate how we will need to develop precise ethical limits for many different domains.

Technical challenges

Requirement 1: Training data sets must measure the intended variables. The fundamental problem here is that the ground truth of whether an individual committed a crime is not available, and can only be estimated via imperfect proxies such as crime reports or arrests. The temptation, then, is to predict these proxies. However, this will likely perpetuate the racial and other biases of the historical system in which these proxies were captured.

Requirement 2: Bias in statistical models must be measured and mitigated. This is easier said than done. There are two widely held misconceptions about bias. The first is that models are only biased if the data they were trained with was inaccurate or incomplete. A second is that predictions can be made unbiased by avoiding the use of protected variables such as race or gender. Both of these ideas are incorrect.

Models can be biased even when the data is accurate and complete. One reason for this is that machine-learning models identify correlation, not causation. For instance, a risk model might identify the number of an offender's criminal friends as a variable that correlates with their probability of reoffending after being released. This will result in a model that is biased against sociable people. As for excluding protected variables, it is all too easy to include other features in the model that are correlated with the protected variable. Recall, for example, the correlation between ZIP code and race.

Requirement 3: Tools must not conflate multiple distinct predictions. Many risk-assessment tools provide composite scores that combine predictions of different outcomes. For example, tools might combine a score measuring a defendant's risk of failing to appear for a scheduled court date with a score measuring the risk of reoffending. However, different causal mechanisms are behind different outcomes. Conflating them into a single model is a recipe for injustice.

Interface issues

Requirement 4: Predictions and how they are made must be easily interpretable. Risk-assessment tools need to be easily interpreted by those who use them. Unfortunately, judges and lawyers often have limited mathematical and technical skills. This requirement is therefore an especial challenge. On top of that, many AI tools are, at least today, black boxes that are hard to interpret, even by experts.

Requirement 5: Tools should produce confidence estimates for their predictions. Risk assessment is not an exact science. The uncertainty of a prediction is an important aspect of any model. In order for users of risk-assessment tools to interpret the results appropriately and correctly, tools should report error bars, confidence intervals or other similar indicators of reliability. If they cannot be provided, the risk-assessment tool should not be used.

Requirement 6: Users of risk-assessment tools must attend trainings on the nature and limitations of the tools. Because of the uncertainty and potential inaccuracies in assessing risk, users of risk-assessment tools should receive rigorous and regular training on interpreting the output. For instance, the training needs to explain how to interpret risk classifications such as quantitative scores or more qualitative 'low/medium/high' ratings. These trainings should address the limitations of the tool.

Governance, transparency and accountability issues

Requirement 7: Policymakers must ensure that public policy goals are appropriately reflected in these tools. Risk-assessment tools implement policy choices, such as the nature and definition of protected categories and how they are used. They should, therefore, be designed and developed carefully to ensure that they achieve desired goals, like the incarceration of only those who present a sufficient risk to society.

Requirement 8: Tool designs, architectures and training data must be open to research, review and criticism. Risk-assessment tools should be as open and transparent as any law, regulation or rule of court. This should prohibit the use of any proprietary risk-assessment tools that rely on claims of trade secrets to prevent transparency. Although Northpointe has refused to reveal how COMPAS works on these grounds, it continues to be widely used. Similarly, training set data needs to available so that independent bodies can check their performance.

Requirement 9: Tools must support data retention and reproducibility to enable meaning contestation and challenges. Justice requires that defendants are able to contest decisions made by risk-assessment tools. For this reason, a defendant should be able to contest the predictions by having access to information about how a tool's predictions are made, such as audit and data trails.

Requirement 10: Jurisdictions must take responsibility for the post-deployment evaluation, monitoring and auditing of these tools. It is not adequate to depend on pre-deployment evaluation of a tool. Crime, law enforcement and justice all have a local flavour. Any jurisdiction using a risk-assessment tool should periodically publish its own independent review, algorithmic impact evaluation or audit of its risk-assessment tools.

If we are going to use risk-assessment tools, then these ten requirements proposed by the Partnership on AI are probably a good place to start. However, I seriously doubt we should go down this road, and so should you. Would you want to hand over your liberty to the decision of a cold and clinical algorithm? Or would you prefer a warm and friendly human judge, despite all their foibles and failings?

One halfway house might be to limit the use of risk-assessment tools to help decide who to release; that is, not to use them to decide who to lock up. Risk-assessment tools might then help enhance, not hurt human rights. There are many other settings where we might safely use AI in such a one-sided way. On the battlefield, for example, we might give weapons autonomy to *disengage* from a target but never to *engage*. We would then secure many of the benefits that the proponents of autonomous weapons have put forward. For instance, Professor Ronald Arkin has argued that we are morally obliged to use autonomous weapons as they will reduce civilian casualties.[13] We can still have this benefit if the weapon can only act autonomously to disengage – when it identifies that a target is civilian and not military, for instance.

ALEXA IS RACIST

In many other settings besides risk assessment, we have examples of algorithms that are racist. Indeed, it is hard to think of a subfield of AI which hasn't had a scandal around racial bias. Take, for example, speech recognition. There has been astounding progress in the capabilities of speech-recognition systems in recent years.

A few decades ago, the idea that speech-recognition systems could be 'speaker independent' was unthinkable. You had to spend hours training a system in a quiet environment and with a high-quality microphone. But we now routinely open a smartphone app and expect it to recognise our voice in a noisy room or on a busy street with little or no training.

If you're Black, though, you will need to dial back your expectations significantly. A 2020 study looked at five state-of-the-art speech-recognition systems developed by Amazon, Apple, Google,

167

IBM and Microsoft.[14] All five systems performed significantly worse for Black speakers than for white speakers. The average word error rate[15] across the five systems was 35 per cent for Black speakers, compared with just 19 per cent for white speakers. The worst-performing system was that provided by Amazon: its word error rate for white people was 22 per cent, but it was over twice as bad for Black people at 45 per cent.

This is not acceptable. If customer service for a major bank or a government welfare agency struggled to understand Black people on the telephone, there would be outrage. If taxis ordered by Black people went to the wrong address twice as often than for white people, there would be many voices calling for the problem to be fixed. We should not tolerate speech-recognition software that is so racially biased.

Other subfields of AI have also suffered from examples of racial bias. For example, computer-vision software still struggles with Black people. I've mentioned Joy Buolamwini's important work uncovering racial biases in facial-recognition software. Then there's the famous Google Photos fail. In 2015, Jacky Alciné found that Google's computer-vision software was tagging pictures of him and his girlfriend as gorillas. His tweet succinctly described the problem: 'Google Photos, y'all f*cked up. My friend's not a gorilla.' There was no easy fix, other than for Google to remove the 'gorilla' tag altogether. We don't know what was behind this fail. It might be biased data. Or it might be more fundamental. AI programs, especially neural networks, are brittle and break in ways that humans don't.

Less well known is that Google Photos will also tag white people as seals. When you or I label a photograph, we understand that mislabelling a Black person as a gorilla, or a white person as a seal, will likely cause offence – and the former is considerably more offensive. But AI programs have no such common sense.

This highlights one of the key differences between artificial and human intelligence. As humans, our performance on tasks often degrades gracefully. But AI systems often break in catastrophic ways. When they are recommending movies on Netflix or ads on Facebook, such fails don't matter greatly. But in high-stakes settings, like sentencing or autonomous warfare, they do.

Racially biased facial-recognition software has already resulted in Black people being wrongly arrested.[16] In 2020, the American Civil Liberties Union (ACLU) filed a formal complaint against Detroit police over what may be the first example of a wrongful arrest caused by faulty facial-recognition technology. Robert Julian-Borchak Williams, a 42-year-old African American man, was arrested after a facial-recognition system identified him incorrectly. The police ran some security footage from a watch store robbery through driving licence records and found a match for Williams. Except it wasn't the correct match, and Williams had an alibi. Even so, the mistake led to Williams spending 30 hours behind bars, as well as experiencing the distress of being arrested at his home, in front of his family.

Racially biased algorithms have also been denying Black people the same healthcare as white people. Two troubling examples have come to light lately. Most recently, a 2020 study of 57,000 people with chronic kidney disease from the Mass General Brigham health system found that Black patients received less access to healthcare.[17] The cause was a racially biased algorithm that produced results in which Black patients were considered healthier than white patients with the same clinical history. In 64 cases, for example, Black patients in the study did not qualify to be placed on the kidney transplant list. However, any one of these 64 patients would have been scored as sick enough to be placed on the transplant list if they had been white.

A second, more subtle case of racial bias in access to healthcare was revealed in a study published in 2019.[18] This second case illustrates again how the use of proxies can lead to racial bias. The racial bias here was in an algorithm used to identify candidates for 'high-risk care management' programs. These care programs provide additional resources for patients with complex health needs, and often result in better health outcomes for these very sick patients. The biased algorithm assigns each patient a risk score that is used in allocating patients to one of these care programs. A patient with a risk score in the 97th percentile and above is automatically enrolled, while a patient with a score in the 55th percentile or above is flagged for possible enrolment, depending on additional input from the patient's doctor.

The study found that Black patients were, on average, far less healthy than white patients assigned the same risk score. As a result, Black patients were significantly less likely to be enrolled in a high-risk care program. The bias arises because the algorithm considers not the health of the patient but a proxy, the health care costs of the patient.

The problem is that we don't have a similar one-dimensional measure of 'health'. The healthcare costs of a patient might seem a simple proxy for the health of a patient. People with higher health-care costs were assigned higher risk scores. But, for various reasons, less is spent on caring for Black patients than for white patients. As with predictive policing, using a proxy for the intended feature embeds a historical racial bias in the output.

Racial bias has had an impact on almost every other aspect of AI. As a final example, it has been observed in natural language processing. AI systems that process language pick up stereotypes and prejudices that exist in the corpus on which they are trained. Such systems therefore tend to perpetuate biases against people of colour.

Perhaps we should not be surprised. AI systems will often reflect the biases of the society in which they are constructed. Training a machine-learning system on historical data will inevitably capture the biases of the past. However, AI systems will put such biases on steroids because of the lack of transparency and accountability within such systems. We therefore have much to worry about.

ALEXA IS SEXIST

Sexism is another major issue which has troubled society in the past and continues to do so today. Not surprisingly, sexism has also arisen in many algorithmic settings, especially those using machine learning. As I mentioned before, the fact that there's a 'sea of dudes' developing AI solutions doesn't help. But the problems go much deeper than this.

Some of the problems are simple and would be easy to fix. Why are voice assistants like Alexa, Siri and Cortana named after women and always speak in female voices by default? The preference for female voices can be traced back to research in military aircraft in the 1980s which suggested that female voices were more likely to get the attention of the mostly male pilots. Some have also argued that women's voices are easier to comprehend. For example, experiments have shown that a baby in the womb is able to pick out their mother's voice. But we're no longer in the womb. And most us have never been in the cockpit of a military jet. Even if this wasn't the case, we could and should avoid the problem altogether.

Names could be gender neutral. Better still, names could inform us that this is a device, not a person. Why not use a name like Alpha or Alif? Voices could be neither obviously male nor obviously female. They might be an intermediate pitch between male and female. And why not make them unmistakenly robotic, so there is no chance they'd be confused with a human?

Some of the examples of sexist algorithms that have emerged are more complex. For example, in 2019 an algorithm deciding credit limits for the new Apple credit card was found to be disappointingly sexist. Apple co-founder Steve Wozniak tweeted that his wife was offered one-tenth the credit limit he was, despite the two sharing the same assets and accounts. The New York Department of Financial Services promptly announced that it would conduct an investigation to see whether the card violated state laws banning sex discrimination. Disappointingly, it was determined that there had not been a violation of the strict rules created to protect women and minorities from lending discrimination. But there was, nevertheless, significant egg on Apple's face as a result. And there's no fundamental reason why the algorithm needed to be sexist. If the developers had tried harder from the outset, they could have easily avoided the problem.

There are other settings where problems are more subtle and insidious. One area where there's perhaps too little concern is algorithmic matching in online dating. In the United States, meeting other people via the internet is now the most popular way for couples to get together.[19] There are the mainstream apps like Bumble, Tinder, OKCupid, Happn, Her, Match, eharmony and Plenty of Fish. But there's an app for every taste. Try SaladMatch if meeting someone who shares your taste in salad is important. Or Bristlr if beards are your thing – obviously. Try GlutenFreeSingles for those with celiac disease. And Amish Dating for the select few Amish people who use a smartphone.

At the heart of all these apps is an algorithm that selects potential matches. And it is worth considering for a moment the long-term generational impact that biases in these algorithms might have. Perhaps they'll encourage more people with gluten intolerance to marry each other. Or more tall men to marry tall women. Whatever it is, we'll see these biases reflected in their children. Will we see a rise in celiac disease as a consequence? Or an increase in the number of very tall people? Algorithms may be slowly changing who we are. There's a lot to be said for the somewhat random method of meeting someone in a bar, or through a friend of a friend.

YOUR COMPUTER BOSS

You may not realise it yet, but computers are often already your boss. If you work for Uber,[20] then an algorithm will decide what you do. Similarly, if you pick items in a warehouse for Amazon, then automated systems are tracking your productivity. Amazon's systems generate warning notices and termination letters without input from a human supervisor. Receive six warnings within a rolling 12-month period and the system automatically fires you.

Handing over power to automated systems in this way is not without significant risks. Amazon is frequently criticised for treating its employees like robots. And it has somehow managed to prevent its 1 million US employees from being protected by a labour union since the company was founded in 1994. Uber has also been frequently criticised for the ways its algorithms treat its drivers; some of whom have resorted to sleeping in their cars. And a number have, sadly, committed suicide due to the financial pressures they experience.

Even for those of us who don't work for Amazon or Uber, algorithms may soon be our boss. One of the most troubling areas in which we see the rise of algorithms is in human resources. In particular, companies are turning to algorithms to help them sift through mountains of CVs and decide who to hire. This is a class action lawsuit waiting to happen.

Amazon spent millions of dollars trying to build just such a tool.[21] A team of about a dozen people in the company's Edinburgh engineering hub used machine learning to scan CVs and identify the most promising candidates. Gender was, of course, not one of the inputs. In most countries, it is prohibited by law to base hiring decisions on gender. But the tool learned other, subtler biases. It favoured people who described themselves using verbs more commonly found on male résumés, such as 'executed' and 'captured'.

Another problem comes from using historical data. Suppose we train the tool using the CVs of people in the company who have been promoted. Then the tool will learn the biases of that system. Perhaps men were promoted more often than women? The tool will learn and replicate this bias.

After several years of effort, in 2018 Amazon gave up and scrapped the idea. Despite the significant machine-learning expertise within the company, they were unable to make it work. It was irretrievably sexist, biased against hiring women.

Ultimately, it is worth noting, there is no unbiased answer. An old-fashioned name for machine learning is 'inductive bias'. We are selecting a small number of people to consider hiring or promoting out of a larger number. This is a bias. We want this decision to be biased in whatever way society finds acceptable. Do we favour those with the best exam results? In other words, is the algorithm biased towards good exam results? But then too many jobs might go to people who went to private schools, which tend to produce better exam results. How do we balance achievement against opportunity? Should we bias the output towards opportunity, and possibly against achievement?

Encapsulating such decisions in algorithms requires us to be much more precise about what sort of society we want. What does it mean, exactly, to be fair? These are not new questions. But we do need to be mathematically prescriptive in our answers if we are to entrust algorithms with such decisions.

INSURING FAIRNESS

AI will throw up fairness questions that fundamentally challenge how many markets operate. All markets require rules to ensure they operate fairly and efficiently. Rules prevent insiders from trading on privileged information, the big guy from exploiting the little one, and externalities from being improperly priced.

Take, for instance, the insurance market. This is essentially about finding a fair price for risk. The challenge is that AI can help us price risk more precisely. In the past, we might have had no easy way to differentiate between customers, and so priced their risk identically. But AI offers the ability to target customers individually.

There are two ways in which AI helps compute risk in a more targeted fashion. First, AI lets us collect new data on customers. For example, car insurance companies can capture tracking data from our smartphones that record how well (or badly) we drive. Second, AI lets us analyse data sets that are too large for human eyes. Health insurance companies can, for instance, mine genetic data and identify those of us at risk of particular types of cancer.

The fundamental tension here is that insurance is about spreading risk over a group, not individual risk. If we price risk individually, there is little point for insurance. We might as well insure ourselves without the overhead of an insurance company taking profit from the deal. The only value left in having insurance is against 'acts of God' that are too catastrophic for any one individual to bear.

Insurance markets reflect society's values about fairness. For example, from 21 December 2012, the European Union mandated that insurance companies charge the same price to men and women for the same insurance products. The policy applies to all forms of insurance, including car insurance, life insurance and annuities. What you pay for insurance should be the same if you are a man or a woman.

As a direct result of this regulation, car insurance rates for women went up. Women now had to subsidise men's poor driving. Accident statistics demonstrate clearly that women are better drivers than men. Three times the number of men die in road traffic accidents compared to women. Is it fair, then, for women to be hit with higher premiums to pay for car accidents caused by too much male testosterone?

AI-based insurance products will raise many fairness questions like this in the near future. They call into question what is a fair and just society. Is it one in which women subsidise men's poor driving? Or in which those with a genetic predisposition to bowel cancer have more expensive health and life insurance? There are no easy answers to such questions.

ALGORITHMIC FAIRNESS

Ultimately, one of the promises of algorithms is that they can make decision-making fairer. Let's not forget that humans are terrible at making unbiased decisions. We like to think that we can make fair decisions, but psychologists and behavioural economists have identified a large catalogue of cognitive biases, systematic deviations that humans make from rational decision-making.

Let me name just a few of the cognitive biases that you and I have, which you may or may not have heard about. Anchoring, belief bias, confirmation bias, distinction bias, the endowment effect, the framing effect, the gambler's fallacy, hindsight bias, information bias, loss aversion, normalcy bias, omission bias, present bias, the recency illusion, systematic bias, risk compensation, selection bias, time-saving bias, unit bias and zero-sum bias. Our poor decision-making runs the gauntlet from A to Z.

One of my favourites is the IKEA effect.[22] This is when we place a disproportionately high value on products that we have ourselves helped create. Having just spent several painful and frustrating hours assembling a Malm chest of drawers, I can confirm that I value it way more now than the $100 I paid for it.

Algorithms offer the promise of defeating all these cognitive biases, of making perfectly rational, fair and evidence-based decisions. Indeed, they even offer the promise of making decisions in settings either where humans are incompetent, such as decisions that require calculating precise conditional probabilities, or where humans are incapable, such as decisions based on data sets of a scale beyond human comprehension.

Unfortunately, the reality is that algorithms have done depressingly little of this superior decision-making yet. I spent a few dispiriting weeks asking AI colleagues for examples in which algorithms had not simply replaced human decision-making but improved upon its fairness. There were far fewer examples than I had hoped for.

The one example that many of my colleagues did mention was the National Resident Matching Program. This is a non-profit organisation in the United States created in 1952 that matches medical students with training programs in teaching hospitals. Algorithms here have made the process of matching students to hospitals fairer for the students.

In 1995, concern arose within the medical community that the algorithm that had been in use for many years to match medical students with hospitals favoured the hospitals over the students. Professor Alvin Roth of Stanford University, who would go on to win the Nobel Prize in Economics for his work in this area, proposed simply switching around the inputs to the algorithm so it favoured the students.[23]

The impact of switching to this 'fairer' algorithm was more theoretical than practical. In practice, the matchings produced by the two algorithms are almost identical: fewer than one in 1000 applicants receive a different match. On the plus side, most (but not all) of the few applicants who are matched to different positions by the new algorithm do better. Nevertheless, the change was very important in restoring the trust of the medical community in the matching system.

THE FUTURE

Despite the limited number of algorithms in everyday use that have increased the fairness of decision-making, I remain optimistic about our algorithmic future. We can expect more and more decisions to be handed over to algorithms. And, if carefully designed, these algorithms will be as fair as, if not fairer, than humans at these tasks. Equally, they'll be many settings where algorithms will be used to make decisions that humans just couldn't process, and do so fairly. Let me end by giving two examples of this promising future.

The first example comes from one of my colleagues, Professor Ariel Procaccia at Harvard University. He has set up a website to help people divide things fairly. I encourage you to check out spliddit.org. The website offers a free service to enable people to divide up the rent in a share house, to split a taxi fare, to assign credit for a group exercise, to divide up the goods in an inheritance or divorce settlement, or to allocate people to a set of chores.

The website uses a bunch of sophisticated algorithms to ensure that the divisions reflect people's preferences and are provably fair. For example, it will divide up the rooms and rent in a share house so that no one thinks anyone else got a better deal. And it will divide up the items in a divorce so no partner envies the other. Can you imagine – no fighting over who gets the battered copy of *The Dice Man*?

If you will indulge me, the second example of algorithms increasing fairness comes from my own work. Back in the 1990s, I helped out with timetabling the exams at Edinburgh University. By getting a computer to do the scheduling, we were able to make better decisions that were fairer on the students than the previous human-generated schedule.

Obviously, when you timetable exams, you can't have a student sit two exams simultaneously. But we were able to go one better. We used the superior horsepower of the computer to come up with schedules where students always got a gap between exams. They never had to sit exams in consecutive slots of the timetable. This is a small but, I hope, persuasive example of how we can expect computers to make better, even fairer decisions for us.

*

Let me combine the lessons that have been sprinkled throughout this chapter. If you are responsibile for building algorithms that will make decisions which have an impact on people, you may wish to keep this list close to hand. Perhaps you are building an algorithm to schedule vaccinations, or a dating website. Before you release your creation, you might want to consider these 12 lessons.

Lesson #1: AI won't fix systems that are essentially unfair. Indeed, AI will often expose and magnify fundamental flaws of unfair systems.

Lesson #2: Be cautious about developing AI systems that take away even part of people's agency.

Lesson #3: Algorithms will be blamed in settings where humans have and should be taking responsibility.

Lesson #4: Using machine learning to make predictions based on historical data is likely to perpetuate the biases of the past. It may even create feedback loops that amplify those biases.

Lesson #5: Be wary of machine-learning systems where we lack the ground truth and make predictions based on some proxy for this.

Lesson #6: Do not confuse correlation with causation. AI systems built with this confusion may perpetuate systemic injustices within society.

Lesson #7: Fairness means many different things, and not all of them can be achieved at the same time. Trade-offs will often be required.

Lesson #8: Limit AI systems to decisions that can only increase and not decrease human rights.

Lesson #9: AI systems will often reproduce and sometimes even amplify the biases of the society in which they are built, such as racism and sexism.

Lesson #10: There are many settings in which there is no unbiased answer. AI systems need, then, to encode what is acceptable to society.

Lesson #11: AI systems will create new markets in which we will need to decide as a society what is fair and just.

Lesson #12: Just building AI systems to match the fairness of human decision-making is a tall order. Making them fairer than humans will be even more difficult.

PRIVACY

Along with fairness, the other area where artificial intelligence has thrown up some of the most pressing ethical questions is privacy. This isn't because AI creates many fresh ethical challenges concerning privacy. Privacy is a long standing and fundamental right that has preoccupied society for many years.

Article 12 of the Universal Declaration of Human Rights, adopted by the UN General Assembly on 10 December 1948, declares: 'No one shall be subjected to arbitrary interference with his privacy, family, home or correspondence, nor to attacks upon his honour and reputation. Everyone has the right to the protection of the law against such interference or attacks.'

AI does, however, put many existing concerns around privacy on steroids. Many of the problems are of what I call the 'boiling frog' form. It is often said that if you put a frog in a pan of boiling water, it jumps straight out. But if you put it in a pan of cold water and slowly increase the temperature, then it will not notice the increasing water temperature and will sit tight until it is too late. This may be a myth, but it is a good metaphor for many privacy problems.

One of the best examples of such a privacy problem is closed-circuit television (CCTV). We've got used to the idea that CCTV cameras are on every street corner. And we haven't seen them as too much of a threat to our privacy, perhaps as we know there are just too many cameras out there to be watched simultaneously. If a crime takes place, then the police will collect the footage from the CCTV cameras and try to identify the perpetrators. This is perhaps not a great invasion of our privacy, as it's after the event. And we know that a crime is being investigated and criminals are being tracked down.

But we can now put some computer-vision algorithms behind all those CCTV cameras. In this way, we can surveil large numbers of people in real time. And we can do this across a whole city, or even a nation. China appears to be building such a system – chillingly called Skynet[1] – which supposedly can scan billions of faces every minute.

George Orwell got one thing wrong in *Nineteen Eighty-Four*. It's not Big Brother watching people. It's not people watching people. It's computers watching people. And computers can do what people cannot do. In 2018, Chinese police used facial-recognition technology to identify and arrest a man at a rock concert with an audience of some 60,000 people.[2] Human eyes can't pick out a person in a crowd of 60,000 people, but computer eyes can.

East Germany demonstrated what was perhaps the limit of human surveillance. At one time, 2 million people – around one in nine of the population – were informants for the Stasi, the state secret police. With Teutonic efficiency, the Stasi collected vast paper files on the East German people. But there's only so far you can go with such human-powered systems.

The beauty of digital technologies is how easy and cheap it is to scale them. It is possible to surveil a whole population using technologies like AI. You can listen to every phone call, read every email, monitor every CCTV camera. And if we look to countries like China, we see that this is starting to happen. That's what I mean by AI putting privacy issues on steroids.

THE HISTORY OF PRIVACY

Vint Cerf is Google's Chief Internet Evangelist. He didn't choose the job title himself, even though Google is the sort of company where you can. Vint had asked if he could be called Archduke. But being called an archduke probably hasn't been such a good idea since the assassination of Archduke Franz Ferdinand. As a result, Vint ended up as Google's Chief Internet Evangelist.

Google hired Vint to advance the internet, set new standards and the like. Chief Internet Evangelist, therefore, is probably a better summary of what he does than Archduke. In any case, Vint is uniquely placed to pursue this role as he is (a) a nice guy and (b) one of the original architects of the internet.[3]

In 2013, he told the US Federal Trade Commission that 'privacy may actually be an anomaly'. He justified this claim with the observation that 'privacy is something which has emerged out of the urban boom coming from the industrial revolution'. Many in Silicon Valley would, I suspect, agree with Vint's remarks. They certainly seem to be acting in a way that ensures privacy becomes an anomaly.

Philosophers, on the other hand, would likely disagree. We've been worrying about privacy for thousands of years. Aristotle, for example, drew the distinction between the public sphere of politics and the private sphere of domestic life. Biologists and anthropologists would also likely disagree. Other animals besides humans seek out privacy. And privacy appears to play an important role within human society, providing, for example, space for contemplation, dissent and change.

On the other hand, from a legal perspective at least, there is something to be said for Vint's claims. Privacy only really entered the US legal system around the time of the Industrial Revolution. A seminal article from 1890 by Samuel Warren and Louis Brandeis in the *Harvard Law Review*[4] first put forward the argument that new technologies which were emerging at that time and invading people's personal space – specifically photography and mass-market newspapers – necessitated new rights, especially the right to privacy.

Warren and Brandeis proposed a number of important legal ideas in this article that live on with us today. For example, they argued that claiming that the private information being published is true is not an acceptable defence against the charge of invading someone's privacy. Similarly, they argued that it is not an acceptable defence to claim simply an absence of malice in publishing such private information.

However, nowhere in their article did Warren and Brandeis suggest that the invention of even newer technologies would or even could undermine these arguments. There is no logical reason to suppose they would. The right to privacy is a response to the ability of new technology to surveil us without our consent. The internet and other digital technologies like AI only increase the necessity to provide such protections.

PRIVACY AND TECHNOLOGY

Some years ago, the Association of the Bar of the City of New York organised a Special Committee on Science and Law to examine the impact of new technologies on privacy. Professor Alan Westin of Columbia University chaired the committee and summarised the committee's findings in an important and influential book titled *Privacy and Freedom*.[5]

The book provides a good summary of many of the privacy problems that new technologies create. It has a US focus, but the concerns it raises are universal. The book begins:

To its profound distress, the American public has recently learned of a revolution in the techniques by which public and private authorities can conduct scientific surveillance over the individual ... As examples mount of the uses made of the new technology, worried protests against 'Big Brother' have set alarms ringing among the civic-group spectrum from extreme left to radical right ... The real need is to move from public awareness of the problem to a sensitive discussion of what can be done to protect privacy in an age when so many forces of science, technology, environment, and society press against it from all sides.

And it ends with some sobering advice:

> American society now seems ready to face the impact of science
> on privacy. Failure to do so would be to leave the foundation of
> our free society in peril. The problem is not just an American one.
> Our science and our social development have made us the first
> modern nation to undergo the crisis of surveillance technology,
> but the other nations of the West are not far behind ... the choices
> are as old as man's history on the planet. Will the tools be used for
> man's liberation or his subjugation? ... [C]an we preserve the
> opportunities for privacy without which our whole system of civil
> liberties may become formalistic ritual? Science and privacy:
> together they constitute twin conditions of freedom in the twen-
> tieth century.

But why the twentieth century? The book appears to be talking
about today, and we're now in the twenty-first century.

What I have avoided telling you is that *Privacy and Freedom*
was written in 1967. And the surveillance technologies discussed
in the book were technologies like miniature cameras and tape
recorders, along with wire taps and polygraphs.

Concerns about the impact of new technologies on our privacy,
then, are neither new nor original. They can be traced back through
Westin's Special Committee in the 1960s to Warren and Brandeis
in 1890, and from there back through many centuries to Aristotle.
And if there's one warning to take from this history lesson, it is
that we are always playing catch-up in our attempts to protect
people's privacy against the encroachment of new technologies.

PREDICTING THE FUTURE

There's a useful maxim often attributed to Hal Varian, the chief economist for Google, but actually coined by Andrew McAfee, the co-founder and co-director of the MIT Initiative on the Digital Economy: 'A simple way to forecast the future is to look at what rich people have today; middle-income people will have something equivalent in 10 years, and poor people will have it in an additional decade.'

For example, rich people today have personal assistants to help them manage their busy lives. In the future, the rest of us will be able to call upon digital assistants to help us manage our lives. Today, rich people have personal chauffeurs to drive them around. In the future, the rest of us will be driven around in autonomous cars. Rich people today have personal bankers to manage their many assets. In the future, the rest of us will have robot bankers that manage our finances. Today, rich people have personal doctors to keep them fit and healthy. In the future, smartwatches and other devices will monitor our bodies and provide advice to keep the rest of us fit and healthy too.

McAfee's rule for predicting the future would seem to suggest that all of us will gain the sort of privacy that today only the rich can afford. Rich people can hide away from the public in VIP suites, fly in personal jets rather than slumming it in economy, and employ expensive lawyers to keep their names out of the papers. I suspect, however, that privacy is the one thing rich people have today that the rest of us won't have in a decade or two. I can try to keep my name out of the newspaper by getting myself a robo-lawyer.

But I fear my robo-lawyer will be no match for the more expensive robo-lawyers that the newspapers will have to keep pesky poor people like me at bay.

More fundamentally, most of the 'free' digital services that we all use today – like social media and search – are paid for with our data. The maxim that 'if you're not paying, you're the product' explains many of the business models of Silicon Valley. Surveillance capitalism depends on exploiting the vast troves of personal data collected by companies like Google and Facebook.

Imagine for a moment a better world in which we could force the technology companies to pay us for our personal data. The problem is that we wouldn't get very much back. In 2019, for example, Google's parent company, Alphabet, made a profit of $34 billion on a turnover of $161 billion. Not bad. But divide this profit among the 5 billion or so people who use Google's search engine and we only get $7 each. That's not going to buy any of us much privacy.

Technology companies like Google can scale their digital services at very little or no financial cost to us. But only because they (and we) don't properly value our personal data. On the other hand, it is not impossible to imagine a better future. If each of us using Google was willing to pay just $30 per year, then Alphabet would have as much income as they get today from advertisers. Then they wouldn't need to sell our data to *anyone*.

INTRUSIVE PLATFORMS

Is such a future really possible? Could we build digital platforms that didn't exploit our personal data without our consent? Could we perhaps build a search engine that didn't construct profiles of its users, and that showed every user the same search results? Actually, you can. It's called DuckDuckGo. It's even free to use. Go check it out.

How about a social network that puts the privacy of its users at the centre of its design? Imagine a social network where the default is that your posts are only visible to people you actually know in the real world. Imagine a social network that doesn't track you with cookies, and that promises it never will. Imagine a social network that lets its users vote on any changes to its privacy policies.

It's hard to remember, but such a social network did actually once exist. It was called Facebook. Unfortunately, over the last decade, Facebook has backtracked on all these good behaviours. When faced with any choice, it seems that the company always choses profit over privacy.

Facebook has achieved its dominance through a long-term strategy of anti-competitive tactics. Buying rivals like Instagram and WhatsApp hasn't helped the consumer. But it has permitted Facebook to continue to grow despite offering worse and worse privacy. Not surprisingly, recent anti-trust action against Facebook has focused on privacy concerns.

Facebook was actually a latecomer to the social networking market. From 2005 to 2008, MySpace was the largest social networking site in the world, with more than 100 million users per month. In June 2006 it was visited by more users than any other website in the United States. On MySpace, the default was that anyone could see your profile.

Facebook distinguished itself by respecting your privacy better than MySpace. It was perceived as a 'safer' space than MySpace. Unlike MySpace, Facebook profiles could be seen only by your friends or people at the same university. Facebook even said: 'We do not and will not use cookies to collect private information from any user' in an early promise, now broken.

When Facebook announced in 2014 that it was acquiring WhatsApp, it declared: 'WhatsApp will operate as a separate company and will honor its commitments to privacy and security.' Naturally, for a messaging service that differentiated itself by offering end-to-end encryption of users' messages, WhatsApp had a much better privacy policy than Facebook back then.

But just 18 months after Facebook's acquisition, WhatsApp started sharing information with Facebook unless users opted out. And in 2021 Facebook made it impossible for WhatsApp users not to share their information with Facebook. That is, unless they happened to be in Europe or the United Kingdom, where stricter data-protection laws apply and even Facebook's expensive lawyers can't ignore them.

If you want to continue to use WhatsApp, you have no choice but to accept the new, weaker privacy policy, one that violates the promises Facebook made when it bought WhatsApp.

In 2018, WhatsApp co-founder Brian Acton left the company, walking away from a cool $850 million of stock options, after disagreeing with Facebook's changes. He told *Forbes* magazine: 'I sold my users' privacy to a larger benefit. I made a choice and a compromise. I live with that every day.' Earlier that year, he had tweeted:

'It is time. #deletefacebook.'[6]

Privacy, or rather the lack of it, was also at the heart of the Facebook and Cambridge Analytica scandal. Facebook knowingly let third parties collect private data on individuals without their consent. In 2019, the US Federal Trade Commission issued Facebook with a record-breaking $5-billion fine as a consequence of these privacy violations.

The penalty was the largest imposed on any company for violating consumers' privacy. It was almost 20 times greater than any previous privacy or data security penalty previously imposed. It was, in fact, one of the largest fines ever given out by the US government. But for Facebook – which had revenue in 2019 of over $70 billion, an annual profit of $18 billion and nearly 2 billion daily active users – the FTC's fine was just a speed bump on the path to global domination. It seems it pays to treat your customers as dumb f—ks.[7]

FACE RECOGNITION

A dozen years ago, I could tolerate my AI colleagues working on face recognition. There were enough positive benefits, it seemed, to allay my fears about developing a technology that would enable surveillance of a nation. On top of this, facial-recognition software, like most AI technologies in the 2000s, was so poor that it didn't appear threatening.

I tried a friend's facial-recognition demo. At the time, I had considerably more hair. His demo, as a result, consistently identified me as a woman. 'You smile too much,' he told me. Clearly, facial-recognition software was not a technology back then that you or I needed to worry much about.

A lot has changed since then. The benefits haven't gone away, but the downsides have multiplied considerably. And facial-recognition software has become sufficiently good that it has crossed out of the lab and into our lives. As a consequence, I'm no longer sure we should be working on facial-recognition software.

There are a few good uses of facial-recognition software. In 2018, for example, Delhi police used it to reunite nearly 3000 missing children with their parents in just four days.[8] Fifteen months later, this software had identified over 10,000 missing children.[9] It's hard not to like such a story.

But India's government now has a much more ambitious and troubling plan for facial-recognition software. It wants to build a single, centralised database covering the whole country using images from social media accounts, newspapers, CCTV cameras, passport photos, publicly available pictures and criminal records. In 2020 it started to use the technology to arrest protesters of a new citizenship law that critics say marginalises Muslims.[10]

Such facial-recognition software remains far from perfect. In 2020, news stories started to break of people being falsely arrested and imprisoned due to facial-recognition software.[11] Amazon, Microsoft and IBM all quickly announced they would stop or pause offering this to law-enforcement agencies. However, the major players in the industry – companies such as Vigilant Solutions, Cognitec, NEC and Clearview AI – continue to provide their software to police departments and other government agencies around the world.

Clearview AI has gained considerable notoriety, along with a number of lawsuits, for its use of facial-recognition software. Its founder and CEO, the Australian entrepreneur Hoan Ton-That, seems determined to push the envelope, and to profit from the resulting publicity. As a consequence, Clearview AI has set several dangerous precedents.

The company scraped 3 billion images of faces from publicly accessible sources such as Facebook and Google. These have been used to create a database which Clearview AI has licensed to more than 600 law-enforcement agencies around the world, as well as to a number of private companies, schools and banks. If you use social media, your photograph is probably in their database.

Clearview AI is facing a number of lawsuits about its services. In 2020 alone, four lawsuits were filed against the company for violations of the *California Consumer Privacy Act* and the *Illinois Biometric Information Act* – specifically, for collecting data without consent. Facebook will have to pay $550 million to settle a similar facial-recognition lawsuit in Illinois,[12] so Clearview AI could be in serious financial trouble if, as I and many others expect, it loses any of these lawsuits.

The tech companies have also tried to curtail Clearview AI's objectionable activities. Twitter, LinkedIn and Google all sent it cease-and-desist letters, while Facebook released a statement demanding that Clearview AI stop using images lifted from the social media platform. Apple didn't bother with any letters and simply suspended Clearview AI's developer account.

Even if Clearview AI ends up being shut down, the problems won't go away. The dilemma with facial-recognition software is that it is bad if it works and bad if it doesn't. Fortunately, we are starting to see some pushback against its use. Both local and national governments are hitting the pause button. San Francisco, Boston and several other cities have introduced bans on the use of facial-recognition software. And the *Facial Recognition and Biometric Technology Moratorium Act*, introduced by Democratic lawmakers to the US Congress in June 2020, attempts, as the name on the tin suggests, to impose a moratorium on the use of facial-recognition software.

Professional societies such as the Association for Computing Machinery, along with organisations like Human Rights Watch and the United Nations, have also called for regulation. This is to be applauded. However, we should be careful of calls to prohibit the use of facial-recognition software because of racial and other biases.

When introducing the legislation into the US Congress, Senator Jeff Merkley motivated the *Facial Recognition and Biometric Technology Moratorium Act* with the argument: 'At a time when Americans are demanding that we address systemic racism in law enforcement, the use of facial recognition technology is a step in the wrong direction. Studies show that this technology brings racial discrimination and bias.'[13]

It is, of course, entirely unacceptable to see Black people incarcerated due to a biased algorithm. But we risk shooting ourselves in the foot if we use bias as reason to demand regulation. Calls to regulate facial-recognition software must recognise the harms caused when the software works, as well as the harms caused when it doesn't.

Arthur C. Clarke was one of the most visionary science fiction authors ever. He predicted the use of many technical advances, including telecommunications satellites, GPS, online banking and machine translation. One of his lasting contributions is his First Law: 'When a distinguished but elderly scientist states that something is possible, he is almost certainly right. When he states that something is impossible, he is very probably wrong.' So when I claim that facial-recognition software may one day be less biased than humans, you should realise that this is almost certainly the case. (If you wish to argue that I am not old enough to be an elderly scientist, bear in mind that Clarke defined 'elderly' as just 30 to 40 years old.)

It's worth noting that our ability as humans to recognise other human faces is highly variable, has a significant hereditary component and is often biased towards people of our own race. Beating humans at face recognition, therefore, isn't a tall order. One day, just as with playing chess, reading X-rays or translating spoken Mandarin into written English, computers will easily out-perform humans. And at this point, we don't want to be morally obliged to use facial-recognition software since it will make fewer mistakes than humans do. We must not overlook the many other harms that facial-recognition software may bring into our lives when it works.

When democracy demonstrators took over the airport in Hong Kong, the first thing they did was take down the cameras. They knew these were one of the greatest threats to their ability to fight for change. Previously, if you were in a large crowd protesting, you were anonymous. Now, facial-recognition software can identify you in real time.

We are creating Jeremy Bentham's panopticon, an 'all-seeing' institution in which one guard can watch all the residents without themselves being seen. And the problem with this digital panopticon is that even if no one is actually watching, it will change what you do. Knowing that the technology exists, that someone could be watching, means you will modify your behaviour.

This is the future that George Orwell and others warned us about.

THE 'GAYDAR' MACHINE

Troubling examples of the sort of future we could face can be seen in the work of Dr Michal Kosinski, Associate Professor of Organizational Behavior at Stanford University. It has to be said that Kosinski appears to go out of his way to court controversy. He did his PhD at the University of Cambridge, looking at how Facebook 'likes' could predict people's personalities. His PhD examiner, Dr Aleksandr Kogan, went on to develop the app for Cambridge Analytica that harvested such information from Facebook and used it for nefarious purposes.

In 2018, Kosinski and a co-author, Yilun Wang, made headlines by publishing controversial research claiming that computer-vision algorithms could predict a person's sexuality from a single image of their face.[14] In 2021, Kosinski doubled down by publishing follow-up work claiming that computer-vision algorithms could also predict a person's political orientation, again from just a single image of their face.[15] These are both examples of technologies that are chilling if they work, and chilling if they don't.

Suppose, for a moment, you could predict someone's sexuality from a single image. There are a number of countries where homosexuality remains illegal. In Afghanistan, Brunei, Iran, Mauritania, Nigeria, Saudi Arabia, Somalia and the UAE, homosexuals can even be sentenced to death. There are several other countries where homosexuals face persecution and violence. You have to ask, then, why Wang and Kosinski did their work in the first place. What possible good can come from software that can detect a person's sexuality from an image of their face? It's not hard to imagine such software being put to some awful uses.

Wang and Kosinski argued in response:

> Some people may wonder if such findings should be made public lest they inspire the very application that we are warning against. We share this concern. However, as governments and companies seem to be already deploying face-based classifiers aimed at detecting intimate traits, there is an urgent need for making policymakers, the general public, and gay communities aware of the risks that they might be facing already. Delaying or abandoning the publication of these findings could deprive individuals of the chance to take preventive measures and policymakers the ability to introduce legislation to protect people.

I am one of these people who wonders if the finding should be made public. Will claiming that sexuality can be identified from a single image inspire others to develop software to do so? What preventative measures could individuals possibly take in one of those countries where homosexuality is illegal? Facial surgery or fleeing the country would seem to be their only options to avoid harm.

There remains the troubling question of whether you really can detect sexuality from a person's face. Physiognomy – the 'science' of judging people's personality from their facial characteristics – has a troubled and largely discredited history dating back to ancient China and Greece. And Wang and Kosinski's work has only muddied these waters further.

Given a single image of a face, Wang and Kosinski reported that their classifier could correctly distinguish between homosexual and heterosexual men in 81 per cent of cases, and in 74 per cent of cases for women. By comparison, they reported that human judges achieved somewhat lower accuracy on the same images: 61 per cent for men and 54 per cent for women.

However, Wang and Kosinski's study had so many problems in its design and execution that it is far from clear whether computer-vision software can predict a person's sexuality from a single image of their face with any meaningful accuracy. Closer analysis of the data suggests that the software wasn't picking out facial features but likely other cues, such as hairstyle, lipstick, eyeliner and eyeglasses.

Would you be surprised to discover that lesbians use less eyeliner than straight women? The software appeared to be using this clue. Or that younger gay men are less likely to have serious facial hair than their heterosexual counterparts? Again, the software appeared to be using this clue. Another big clue was whether the person was smiling or not. Homosexual males were more likely to be smiling than heterosexual males, while homosexual women were less likely.

The images themselves were problematic. There were no people of colour. The images were scraped from a dating website without the consent of the participants. And bisexuality and other forms of sexuality were ignored. Surely we've learned by now that sexuality isn't a binary thing. It is not that everyone is simply either heterosexual or homosexual, as the researchers' classifier supposed.

The claim that the software was more accurate than humans at predicting sexuality is also problematic. The human experts were simply random people paid to make predictions on Amazon's Mechanical Turk website. It's not at all clear whether this group of people is any good at the task, or has any incentive to make good predictions.

Finally, Wang and Kosinski's statistics – such as 81 per cent accuracy in predicting gay men from straight men – were calculated from a test set with an equal proportion of homosexual and heterosexual people. What happens instead when you have 7 per cent gay men and 93 per cent straight men, as you would find in the US population as a whole? Accuracy then goes down significantly. In such a sample of 1000 men, of the 100 men predicted most likely to be gay by their classifier, only 47 were actually gay. That means 53 per cent were classified incorrectly. The software doesn't appear so good now.

Irrespective of its merits and its failings, Kosinski did, however, achieve his aim of disturbing us – both with his study and with the future it foretells.

TREES IN THE FOREST

Eric Schmidt is the 'grown-up' who was brought in to be the CEO of Google, in charge of its founders, Larry Page and Sergey Brin. Schmidt famously described Google's product strategy as follows: 'The Google policy on a lot of things is to get right up to the creepy line and not cross it.'[16]

Google went up to that creepy line early in the company's existence. In 2004, when it launched the Gmail service to the public, numerous privacy organisations complained that Google was reading people's email in order to place context-sensitive adverts alongside their emails. This is another one of those privacy problems on steroids. We've been worried about people reading our correspondence for centuries. Cryptography was invented in part to deal with such concerns. And people had been reading other people's emails long before Gmail came along.

Back in the 1980s, when I first started using email, there was a meme doing the rounds that the FBI and other agencies read emails containing trigger words like 'bomb' and 'terrorist'. Some people added signatures to the end of their emails full of such trigger words to overwhelm the supposed human readers of emails. The difference now was that Google had built the infrastructure to do this at scale, and it was very open about what it was doing. Google ignored the privacy complaints that came in about Gmail. They continued to read everyone's emails and serve up adverts.

Of course, no person is technically *reading* the emails. Philosophers have debated such a question. If a tree falls in a forest and no one is around to hear it, does it make a sound? If an algorithm, rather than a conscious mind, is reading your email, is your privacy really invaded?

Despite this philosophical get-out, in May 2013 a class-action lawsuit was brought against Google, claiming it 'unlawfully opens up, reads, and acquires the content of people's private email messages', in violation of wire tap laws. Defending itself against the lawsuit, Google argued that the 425 million users of Gmail had no 'reasonable expectation' that their communications were confidential.[17]

Warning enough, Gmail users?

Google claimed: 'Just as a sender of a letter to a business colleague cannot be surprised that the recipient's assistant opens the letter, people who use web-based email today cannot be surprised if their communications are processed by the recipient's ECS [electronic communications service] provider in the course of delivery.'

Google's analogy is somewhat problematic. I expect the postman to read the address on my postcard; I'm a little disappointed (but perhaps not too surprised) if he goes on to read the message. But I'd be very upset if he opened up my letter, read the contents and then included a flyer for a business that might interest me.

In 2018 the story took a familiar turn for those used to the Silicon Valley playbook. News broke that Google was not just reading emails, but had given access to the contents of emails to third-party apps. Worse still, employees of the developers of these apps had read thousands of emails.[18] If this sounds familiar, giving third-party apps access to private data was precisely what got Facebook in trouble with Cambridge Analytica.

If there's any lesson we can draw from these sorry stories, it is that technology will inevitably get creepy unless we put safeguards in place.

ANALOGUE PRIVACY

The second law of thermodynamics states that the total entropy of a system – the amount of disorder – only ever increases. In other words, the amount of order only ever decreases. Privacy is similar to entropy. Privacy is only ever decreasing. Privacy is not something you can take back. I cannot take back from you the knowledge that I sing Abba songs badly in the shower. Just as you can't take back from me the fact that I found out how you vote.

There are different forms of privacy. There's our digital online privacy, all the information about our lives in cyberspace. You might think our digital privacy is already lost. We have given too much of it to companies like Facebook and Google. Then there's our analogue offline privacy, all the information about our lives in the physical world. Is there hope that we'll keep hold of our analogue privacy?

The problem is that we are connecting ourselves, our homes and our workplaces to lots of internet-enabled devices: smartwatches, smart lightbulbs, toasters, fridges, weighing scales, running machines, doorbells and front door locks. And all these devices are interconnected, carefully recording everything we do. Our location. Our heartbeat. Our blood pressure. Our weight. The smile or frown on our face. Our food intake. Our visits to the toilet. Our workouts.

These devices will monitor us 24/7, and companies like Google and Amazon will collate all this information. Why do you think

Google bought both Nest and Fitbit recently? And why do you think Amazon acquired two smart-home companies, Ring and Blink, and built their own smartwatch? They're in an arms race to know us better.

The benefits to these companies are obvious. The more they know about us, the more they can target us with adverts and products. There's one of Amazon's famous 'flywheels' in this. Many of the products they will sell us will collect more data on us. And that data will help target us to make more purchases.

The benefits to us are also obvious. All this health data can help us be healthier. And our longer lives will be easier, as lights switch on when we enter a room, and thermostats move automatically to our preferred temperature. The better these companies know us, the better their recommendations will be. They'll recommend only movies we want to watch, songs we want to listen to and products we want to buy.

But there are also many potential pitfalls. What if your health insurance premiums increase every time you miss a gym class? Or your fridge orders too much comfort food? Or what if your employer sacks you because your smartwatch reveals you took too many toilet breaks?

With our digital selves, we can pretend to be someone that we are not. We can lie about our preferences. We can connect anonymously using VPNs and fake email accounts. But it is much harder to lie about your analogue self. We have little control over how fast our heart beats or how widely the pupils of our eyes dilate.

We've already seen political parties manipulate how we vote based on our digital footprint. What more could they do if they really understood how we responded physically to their messages? Imagine a political party that could access everyone's heartbeat and blood pressure. Even George Orwell didn't go that far.

Worse still, we are giving this analogue data to private companies that are not very good at sharing their profits with us. When you send your saliva off to 23AndMe for genetic testing, you are giving them access to the core of who you are, your DNA. If 23AndMe happens to use your DNA to develop a cure for a rare genetic disease that you possess, you will probably have to pay for that cure. The 23AndMe terms and conditions make this clear:

> You understand that by providing any sample, having your Genetic Information processed, accessing your Genetic Information, or providing Self-Reported Information, you acquire no rights in any research or commercial products that may be developed by 23andMe or its collaborating partners. You specifically understand that you will not receive compensation for any research or commercial products that include or result from your Genetic Information or Self-Reported Information.

A PRIVATE FUTURE

How, then, might we put safeguards in place to preserve our privacy in an AI-enabled world? I have two simple fixes. One is regulatory and could be implemented today. The other is technological and is something for the future, when we have AI that is smarter and more capable of defending our privacy.

The technology companies all have long terms of service and privacy policies. If you have lots of spare time, you can read them. Researchers at Carnegie Mellon University calculated that the average internet user would have to spend 76 work days each year just to read all the things that they have agreed to online.[19] But what then? If you don't like what you read, what choices do you have?

All you can do today, it seems, is log off and not use their service. You can't demand greater privacy than the technology companies are willing to provide. If you don't like Gmail reading your emails, you can't use Gmail. Worse than that, you'd better not email anyone with a Gmail account, as Google will read any emails that go through the Gmail system.

So, here's a simple alternative. All digital services must provide four changeable levels of privacy.

Level 1: They keep no information about you beyond your username, email and password.

Level 2: They keep information on you to provide you with a better service, but they do not share this information with anyone.

Level 3: They keep information on you that they may share with sister companies.

Level 4: They may consider the information that they collect on you as public.

And you can change the level of privacy with one click from the settings page. And any changes are retrospective, so if you select Level 1 privacy, the company must delete all information they currently have on you, beyond your username, email and password. In addition, there's a requirement that all data beyond Level 1 privacy is deleted after three years unless you opt in explicitly for it to be kept. Think of this as a digital right to be forgotten.

I grew up in the 1970s and 1980s. My many youthful transgressions have, thankfully, been lost in the mists of time. They will not haunt me when I apply for a new job or run for political office. I fear, however, for young people today, whose every post on social media is archived and waiting to be printed off by some prospective employer or political opponent. This is one reason why we need a digital right to be forgotten.

That leaves me with a technological fix. At some point in the future, all our devices will contain AI agents helping to connect us that can also protect our privacy. AI will move from the centre to the edge, away from the cloud and onto our devices. These AI agents will monitor the data entering and leaving our devices. They will do their best to ensure that data about us that we want to keep private isn't shared.

We are perhaps at the technological low point today. To do anything interesting, we need to send data up into the cloud, to tap into the vast computational resources that can be found there. Siri, for instance, doesn't run on your iPhone but on Apple's vast servers. And once your data leaves your possession, you might as well consider it public. But we can look forward to a future where

AI is small enough and smart enough to run on your device itself, and your data never has to be sent anywhere.

This is the sort of AI-enabled future where technology and regulation will not simply help preserve our privacy, but even enhance it.

THE PLANET

There remains one more ethical elephant in the corner of the room that I want to consider. Alongside fairness and privacy, this is a third moral problem that has to be urgently addressed as AI takes on an ever larger role in our lives. Indeed, it is arguably the biggest moral problem facing humanity this century. It is, of course, the climate emergency.

What sort of planet are we leaving for our children? How do we prevent irreversible changes to the climate that will cause catastrophic harms to life on earth? Heatwaves and droughts. Hurricanes and typhoons. Ecosystem collapse and mass extinction. Food insecurity and hunger. Death and disease. Economic pain and ever greater inequality.

It's imperative today that we consider how AI may contribute to this ethical challenge. On the one hand, how can AI help tackle the climate emergency? On the other hand, how might AI exacerbate it?

GREEN AI

In the last few years, it's become fashionable to worry about the terrible amount of energy used by AI algorithms. Here is the dean

of engineering at Australian National University being interviewed in the *Australian Financial Review*: '[T]here's also a strong
interest [within AI] in the Sustainable Development Goals, which
seems incongruous with AI, but at the moment 10 to 20 per cent
of the world's energy is consumed by data.'[1]

This is just plain wrong. Less than 20 per cent of all world
energy consumption is electricity. And less than 5 per cent of all
electricity is used to power computers. This means that less than
1 per cent of world energy consumption is spent on all computing.
And AI makes up a small segment of that. So AI consumes just a
fraction of a percent of the world's energy.

Erroneous memes like this can, I suspect, be traced back to a
widely reported study from 2015.[2] This predicted that data centres
could be consuming half of global energy by 2030, and would be
responsible for a quarter of all greenhouse gases pumped out into
the atmosphere. However, reality seems to be shaping up quite
differently.

Data centres have been getting more efficient faster than they
have been increasing in size, and mostly they've been switching to
green renewable energy. Their carbon footprint hasn't therefore
increased as predicted back in 2015. Indeed, if anything, the footprint of data centres might have decreased slightly.[3] It would help,
of course, if the data centre industry was more transparent about
its green future. But even if we discount their still-to-be-delivered
promises, the sector is doing a relatively good job.

Another perspective comes not from the total footprint of AI
but from the individual. How much energy is used, and carbon

dioxide produced, by a single AI model? This is not an easy question to answer. Machine-learning models, for example, come in many different sizes. In May 2020, OpenAI announced GPT-3: at that time, this was the largest AI model ever built, with an impressive 175 billion parameters.

Training this enormous model is estimated to have produced 85,000 kilograms of CO_2.[4] To put this in perspective, this is the same amount produced by four people flying a round trip from London to Sydney in business class. If you're not so familiar with the front of the aeroplane, this is the same as the amount produced carrying the people in a row of eight economy-class seats for the same trip.

This figure of 85,000 kilograms of CO_2 supposes that the energy produced to train the model comes from conventional power sources. In practice, many data centres run on renewable energy. The three big cloud computing providers are Google Cloud, Microsoft Azure and Amazon Web Services. Google Cloud claims to have 'zero net carbon emissions'. Microsoft Azure runs its cloud on 60 per cent renewable energy and has offset the rest since 2014.[5] Indeed, by 2030, the whole of Microsoft plans to be carbon negative. Amazon Web Services, which is the largest provider with well over one-third of the market, is less green. It has, however, promised to be net zero by 2040. Today, Amazon uses around 50 per cent renewable energy when offsets are factored in.

Even if GPT-3 was trained using energy derived only from fossil fuels, it would not be too expensive to offset the CO_2 produced. I put the numbers into myclimate.org and came up with a rough estimate of how much it would cost: you would need to invest around $3000 in projects in developing and newly industri-alising countries to offset the CO_2 produced by training GPT-3. This may sound a lot, but it would have cost over $4 million to pay for the compute to train GPT-3. That was if Microsoft hadn't donated their cloud to OpenAI as part of their strategic invest-ment in the start-up.

It is also worth noting that a model like GPT-3 is very much an outlier. There are only half a dozen or so research labs on the planet with the resources to build AI models as big as this. DeepMind. Google Brain. Microsoft Research. Facebook Research. IBM. OpenAI. Baidu Research. That's about it. Most AI researchers build models that are thousands of times smaller. And training their models produces only kilograms, not tonnes, of CO_2.

A distinction also needs to be made between training and pre-diction. It costs a lot more to train a model than to use it to make a prediction. Training can take days or even months on thousands or tens of thousands of processors. Prediction, on the other hand, takes milliseconds, often on a single core. The amount of CO_2 pro-duced to make a prediction using even a very large AI model can thus be measured in grams. We should therefore divide the amount of CO_2 produced in training a model by the number of times the model is actually used to make predictions.

ON THE EDGE

I mentioned that AI will start to move from the centre to the edge – that is, away from the cloud and onto our devices. Speech

213

recognition, for example, won't happen up in the cloud, as now, but on our devices. This will help maintain our privacy as our personal data won't have to go anywhere.

This trend of AI moving towards the edge will also help arrest the growing energy demands of AI models. A lot of recent progress in AI has been achieved by increasingly large and computationally intensive deep learning models. Indeed, the demands for computational power to run these ever bigger models is rising more quickly than the power of computers to run these models.

In 1965, Gordon E. Moore, the co-founder of Intel, made a remarkable observation that the number of transistors on a chip doubles about every two years. The number of transistors is a rough measure of the power of a chip. This has become known as Moore's law. It's an empirical law that has held for over half a century, though it is now coming to an end as we reach the quantum limits of how far transistors can be shrunk.

Before 2012, AI research closely followed Moore's law. The compute needed to generate the latest state-of-the-art results was doubled every two years. But as the deep learning revolution took off, the large tech companies especially have been throwing more and more cloud resources at AI in order to make progress. Since 2012, the compute needed to achieve the latest state-of-the-art results has been doubling every 3.4 months.[6] This blistering pace is clearly not sustainable.

The progress in AI since 2012 hasn't just been about compute. The underlying algorithms have also been improving. And the improvement in algorithms has been faster than Moore's law. For example, since 2012, the compute necessary to achieve a particular performance on an AI problem like image recognition has been halving every 16 months.[7] To put it another way, image-recognition methods since 2012 have improved eleven-fold due to hardware, as predicted by Moore's law. But at the same time the software has improved by 44 times due to smarter algorithms.

These algorithmic improvements mean that we will increasingly be able to run AI programs on our devices. We won't have to tap into the vast resources available in the cloud. And because of the small size of our devices, and because of limits in battery technology, we'll do more and more AI with less and less power.

There are plenty of harms that AI will do, but using too much energy is likely to be one of the lesser evils, and one that isn't going to be too hard to fix.

BIG OIL

We should probably be more concerned about the relationship between the Big Tech companies and Big Oil. Silicon Valley has its eyes on Houston. All the tech companies are signing up the oil and gas companies to lucrative deals selling cloud and AI services.

In 2018, Google started an oil and gas division for this purpose. The company has promised that its AI tools and cloud services will enable fossil fuel companies to act better on their data. That is, it will help them extract oil and gas from existing and new reserves faster and more efficiently.

Microsoft is also looking to engage with this sector. In September 2019, the company announced a partnership with the oil and drilling giants Chevron and Schlumberger. Joseph C. Geagea, then the executive vice-president of technology, projects and services at Chevron, said that the collaboration would 'generate new exploration opportunities and bring prospects to development more quickly and with more certainty'.

Amazon Web Services has also been looking to attract the oil and gas industry to its cloud. Until April 2020, AWS had a dedicated website for oil and gas companies, promising to 'accelerate digital transformation, unleash innovation to optimize production and profitability, and improve cost and operational efficiencies necessary to compete under the pressures of today's global energy market'. It has now replaced this with a website that talks more about renewables and sustainability, even if the solutions described continue to focus on exploration and extraction.

Facilitating the extraction of fossil fuels is not something we should encourage. If we are to address the climate emergency before the harms are too great, we must try to keep as much carbon in the ground as possible. Given that the tech companies have plentiful other ways of making money, it hardly seems responsible or wise for them to help speed up the process of finding new fossil fuels to get out of the ground.

There is an important statistic that is perhaps not known widely enough. Just 100 companies are responsible for 71 per cent of global emissions. Since the Intergovernmental Panel on Climate Change was established in 1988, more than half of all industrial emissions can be traced back to the activities of just 25 of these 100 companies. We need to hold these companies to account for their impact on the planet. And we should hold to account any other companies that support these mega-polluters by providing, for example, technological support to continue their operations.

CLIMATE ACTION

Let us consider the other side of the coin. Artificial intelligence can significantly help us tackle the climate emergency. This is good news because we have no time to waste. NASA has reported that the Earth's global average surface temperature in 2020 tied with that in 2016, making these the warmest years on record. And unlike 2016, the temperatures in 2020 were not boosted by a strong El Niño. Of course, whether one year is a record is not really critical. We need to consider longer-term trends. But those trends are not looking good. The last seven years have been the warmest seven years on record. And we see ever-increasing numbers of extreme events, such as bushfires, hurricanes and droughts.

There are many ways that AI already is or will assist us in addressing the climate emergency. Perhaps most directly, it can help us do things more efficiently. If we use and waste less energy, we'll put less CO_2 into the atmosphere. AI can help us both individually and collectively. Smarter appliances in our homes will use less energy and save us money. On the other side, energy companies can leverage vast amounts of data to forecast energy generation and demand, to predict weather patterns, wind and sun so that we can better use renewables.

Another area where artificial intelligence has great potential is transport. Transportation is responsible for around one-quarter of global CO_2 emissions. AI can be a great asset in reducing these emissions.

I've worked with a number of large multinational companies helping them optimise their supply chains. My algorithms solve what mathematicians call 'travelling salesperson problems', finding the best routing for a fleet of trucks. These algorithms can routinely cut transport costs by 10 per cent. The CEOs love that they spend 10 per cent less on transport, as that contributes directly to the bottom line, but that also means 10 per cent fewer kilometres for their trucks to drive, and so 10 per cent less fuel and 10 per cent less CO_2 that goes up into the atmosphere.

Another area where AI can help is in the many forms of production. Take, for example, the production of cement. This alone is responsible for 5 per cent of global greenhouse gas emissions. A study by McKinsey found that AI could improve throughput and reduce energy use in a typical cement plant by up to 10 per cent.[8] What is not to like about reduced CO_2 emissions and increased profits? Steel production, which is responsible for nearly another 5 per cent of global emissions, could benefit similarly from using AI.

Of course, AI alone is not going to be enough to prevent irreversible and destructive changes to the climate. Reductions of 10 per cent here or there are inadequate. We need to get to net zero. And to do so, we must fundamentally change how we live so that we tread more lightly on the planet. We need to eat less meat and more locally produced vegetables. To reduce, reuse and recycle. But AI can perhaps help us keep a few of those bad habits we've acquired in the last hundred years.

AI FOR GOOD

The potential benefits of AI go beyond just addressing the climate emergency. One pleasing aspect in the last decade is how quickly researchers have embraced the idea of artificial intelligence for good.

One of the ethical conundrums that AI creates is that much of it is inherently dual-use. It can be used for good or bad. How, then, do we encourage the good uses and avoid the bad?

The same computer-vision algorithms that go into an autonomous car to track and avoid pedestrians can go into an autonomous drone to track and target people on the ground. We want such tracking algorithms to be developed and refined to make autonomous cars safer. But how do we prevent the very same algorithms being put to more morally challenging uses, such as in kamikaze drones? The same facial-recognition software that identifies a young child lost at a train station can be used to identify and help arrest people exercising their right to protest in a peaceful political demonstration. We might be keen to use such technology to find lost children, but how do we prevent an authoritarian regime from misusing the same software?

For AI researchers, especially in universities, it is hard to limit what people do with our research. We publish in the open. We release our code for anyone to download. We share our results freely with the world, in the hope that it will be picked up and applied. It's hard, then, to prevent our research from being put to uses of which we disapprove. I still recall the moment an industry person came up to me at a conference many years ago to tell me they'd applied my latest scheduling algorithm. It was the first practical application I had come across of my research. The smile fell from my face when I heard it was for Thales' new missile system.

As academics, we have a lot of freedom to choose the applications to which *we* apply our ideas. Therefore, many of my colleagues are turning to those applications of AI which have direct social good. We may not be able to prevent the misuse of our research, but we can at least encourage its positive uses by putting *our effort* into applying it for good.

I often argue that this is one of the reasons academic tenure was invented. If it's a problem that business is going to solve, then I'm not that interested. There'll be money and incentives out there to solve this, I think, so it doesn't need my help. Give me instead a societal problem where no one wants to pay for a solution. This sounds just right for me! Many of my colleagues agree.

'AI for Good' has become a burgeoning sub-area of AI, with many seminars, conference and journals starting up to showcase work in this area. The 17 Sustainable Development Goals (SDGs) put forward by the United Nations in 2015 provide an elegant means to classify most research in this space. The SDGs are a set of interlinked and global goals concerning the societal, economic and environmental wellbeing of the planet. They are designed to be a 'blueprint to achieve a better and more sustainable future for all' by 2030.

The goals have been made more 'actionable' by a 2017 UN Resolution identifying 169 distinct targets. For example, the 2030 health targets include reducing maternal mortality rates globally to less than 70 per 100,000 live births, neonatal mortality to 12 or fewer deaths per 1000 live births, and under-five mortality to 25 or fewer children per 1000 live births. Here, then, is an action plan for a better, kinder, more sustainable planet.

A consensus-based expert elicitation process found that AI might help achieve 134 of these 169 targets.[9] I am struggling to think of another technology that could touch so much of what we need to do by 2030. On the other hand, the consensus of these experts was also that AI might inhibit 59 of the targets. In a crude sense, might we decide that AI can do more than twice as much good as harm?

For instance, predictive policing tools could help achieve the SDG 11, 'making cities safer'. On the other hand, they may hinder the SDG 16, 'justice for all'. As a second example, autonomous vehicles are surely an essential component of achieving one of the 169 targets, which looks to halve the number of global deaths and injuries from road traffic accidents. On the other hand, by eliminating the jobs of many truck and taxi drivers, autonomous vehicles may hinder the achievement of another of the 169 targets: full and productive employment and decent work for all.

I can't see it any other way. There are few technologies which only bring benefits. AI is no exception.

THE WAY AHEAD

MORAL MACHINES

It should be clear by now that we cannot today build moral machines, machines that capture our human values and that can be held accountable for their decisions. And there are many reasons why I suspect we will likely never be able to do so.

Machines will always and only ever be machines. They do not have our moral compass. Only humans can be held accountable for their decisions. Machines do not suffer or feel pain. They cannot be punished. Machines aren't, and probably won't ever be, conscious of their actions. They are made of the wrong stuff to be moral beings.

If we can't build moral machines, it follows that we should limit the decisions that we hand over to machines. But here's the rub: we're not going to be able to avoid handing over *some* of our decisions to machines. And in some cases, it might actually make the world a better place if we did. Indeed, we already have handed over certain decisions to machines.

Many of these are low-stakes decisions that improve our lives. How do you get from A to B? Let your satellite-navigation system decide. What song should you listen to next? Let your smart speaker decide. In such cases, technology only results in us spending less time lost and more time musically satisfied.

But some decisions we hand over to machines will be high-stakes. What grades do we give to students who couldn't sit their exams because of the pandemic? To which prisoners do we give parole? Which taxpayers do we audit? Who should be shortlisted for a job interview?

Even if machines can make high-stakes decisions like these 'better' than humans, we might choose not to give all such decisions to machines. We might continue to have humans make certain high-stakes decisions, despite all the imperfections inherent in human decision-making. It might be better to have human empathy and human accountability despite human fallibility. And this might be preferable to the logic of cold, unaccountable and slightly less fallible machines.

In our courtrooms, perhaps human judges should always decide whether a person will be locked up. Machines could possibly make society safer, only locking up those who truly are a risk to the community. But doing so would transform the world in which we live, turning it into one of the bad dreams envisaged by writers like George Orwell and Aldous Huxley.

This leaves us with a fundamental ethical problem, one that only us humans can address. We cannot build moral machines, but we will let machines make certain decisions of a moral nature. How, then, should we ensure that the world is a better place despite this?

I don't believe we know the answers to this question yet. But I believe we can identify some of the tools – technical, regulatory and educational – that will be needed.

TRUSTING AI

Ultimately, we want to build AI that we can trust. It's the same with any technology. If we're building a nuclear power station, we want to trust that it has safety systems in place that mean it will never melt down. If we're building a new jet plane, we want to trust that it will never fall out of the sky. Similarly, with AI systems, and especially those with autonomy, we want to trust that they'll always do the right thing.

Trust is a complex beast. It's difficult to define and hard to earn. In the context of AI systems, it can be broken down into a number of desirable characteristics. Let me list seven key characteristics of AI systems that might engender trust.

Explainability

A system that can explain its decisions is more likely to be trusted than one that is a black box. Any explanations that a system produces need to be understandable. They have to speak the language of the user, not of the builder of the system. Unfortunately, many AI systems today do not provide meaningful explanations of their decisions. There is, however, lots of research underway in XAI, or 'explainable AI'. Promising progress is being made in several areas, such as counterfactual explanations.

Auditability

When something goes wrong, we need to be able to work out what happened. AI systems therefore need to be auditable. Our trust in aeroplanes improved greatly following the Australian invention of the black box flight recorder. Accidents could then be audited, mistakes identified and systems re-engineered to try to prevent a repeat of the events that led to the accident. As a consequence, commercial flying is now one of the safest forms of transport. Similarly, AI systems will need 'flight recorders' so that they too can be audited when mistakes occur.

Robustness

We trust systems that behave in predictable ways, and that are capable of dealing with perturbations to their inputs. AI systems need, therefore, to work robustly. Unfortunately, many AI systems today are brittle, breaking when the input changes even in small ways. We can invert a single pixel in an image of a gun, and the object classification system will wrongly classify the image as a banana. It is hard to trust systems which fall over so readily.

Correctness

Especially when human lives are at stake, we want very strong assurances that an AI system can be trusted to act correctly. For example, can we prove mathematically that the airplane's fly-by-wire system will never crash? Or that the reactor's water temperature will remain within safe bounds whatever happens? Unfortunately, the open-ended nature of many intelligent tasks may mean that there are limited settings where we can formally prove that a system is correct and does precisely and only what it is supposed to do.

Fairness

We expect systems to treat us fairly. For instance, an AI system that is racist or sexist is not one that people are likely to trust. While there are many examples of AI systems behaving unfairly, they actually have the potential to be fairer than humans. Human decision-making is full of conscious and unconscious biases. AI systems, on the other hand, can be evidence-based and, if carefully programmed, objectively fairer.

Respect for privacy

AI systems work at speeds, scales and costs that mean we can apply them to tasks that threaten people's privacy. Face-recognition systems can surveil every street corner. Speech-recognition software can listen in on every telephone call. And natural language systems can read every email. It is likely that we will distrust systems that invade our privacy, or treat our data carelessly.

Transparency

The seventh and perhaps most overrated characteristic that could help engender trust is transparency. Transparency can be a useful means of helping gain people's trust. However, as I discuss shortly, there are also pitfalls to being transparent. And transparency itself does not necessarily engender trust. WhatsApp's transparency about its intention to start sharing my information with Facebook made me trust it less, not more.

*

We are still working out how to engineer systems with all these characteristics. In fact, we barely know how to build systems which have even a few of them. And trust is not just an engineering issue. We need to consider the broader socio-technical setting in which our AI systems sit.

For instance, how do we create the norms that will improve trust in AI systems? What sort of regulatory environment will help engender trust? How do we prevent automation bias in humans from undermining trust in AI systems? And what sort of explanations from an AI system will humans appreciate and trust?

It's worth noting that the people we trust also often lack many of these characteristics. We are, for example, frequently very poor at explaining ourselves. All of us make biased and unfair decisions. It's not only impossible to prove the correctness of our decision-making, but there's actually plenty of evidence from behavioural psychology that many of our decisions are flawed and incorrect. Finally, despite spectacular advances in our scientific understanding, the human brain is far from transparent.

Yet this is not a double standard. We should hold machines to higher standards than humans. And this is for two important reasons. First, we should hold them to higher standards because machines, unlike humans, are not and likely can never be accountable. We can put up with a lack of transparency in human decision-making because, when things go wrong, we can call people to account, even punish them. And second, we should hold machines to higher standards because we can. We should aspire to improve the quality and reliability of human decision-making too.

TRANSPARENCY

Frequently, transparency is put forward as an important way of engendering trust in AI systems. IBM, for instance, has made transparency a central part of its mission, listing it as one of just three ethical principles that guide its use of AI. But it's not only IBM that has elevated transparency to a position of great importance.

In 2019, for example, the European Commission published its guidelines that AI systems should meet in order to be deemed trustworthy.[1] Transparency was one of just seven key characteristics put forward. Many other organisations have called for transparency within AI systems, from the G20 group of governments to DeepMind and Deutsche Telekom.

Transparency is certainly a useful tool for building trust. But, at best, it is a means to an end. It is not the end in itself. Facebook's transparency report boldly proclaims:

> We're committed to making Facebook a place that's open and authentic, while safeguarding people's private data and keeping our platform safe for everyone. We publish regular reports to give our community visibility into how we enforce policies, respond to data requests and protect intellectual property, while monitoring dynamics that limit access to Facebook products.[2]

Despite this transparency, Facebook is the least trusted of the four Big Tech giants. In a recent survey, just 5 per cent of people trusted the company.[3] I suspect greater transparency from Facebook will only increase the public's distrust.

There are also many situations in which transparency is undesirable. For example, many companies use trade secrecy to protect valuable intellectual property. Google rightly doesn't share the secret sauce in its search algorithm. Not only is this the only way to protect the billions of dollars it invests into improving search, it also helps prevent bad actors from manipulating search results. Transparency in this case would be a bad thing.

There is, however, one area in which we could do with more transparency. This is in informing people where AI is being used. In 2016 I proposed a new law to ensure better transparency in the use of AI.[4] I named it the 'Turing Red Flag law' in honour of Alan Turing, one of the founders of the field of artificial intelligence. I also named it in recognition of the people who would walk in front of cars waving red flags at the dawn of motoring to warn people of the strange contraptions coming down the road. The Turing Red Flag law states: 'An AI system should be designed not to be mistaken for a human.'

It didn't take long to see situations where my new law might be needed. In 2018, for instance, Google brought out Duplex, their new voice assistant. The demo of Duplex at I/O, Google's flagship developer conference in Mountain View, California, in May 2018 stole the show. But it caused concerned headlines around the world.

In the Duplex demo, the computer voice assistant rang up a hairdresser to book an appointment, and then a restaurant to book a table. The voice assistant had a completely lifelike conversation with the person on the other end of the phone. They appeared to have no idea it was a computer, and not a real person, talking to them. To reinforce the confusion, the voice assistant ummed and erred like a real person. I played the demo to many people. Most couldn't work out who was real and who was the computer.

As I told reporters at the time, what possible reason could there be to umm and err other than to deceive? What's worse, I'd been told by people I knew inside Google that management were advised to start any call with a warning that this was a computer and not a person, but they chose to ignore this advice. Few saw the irony that CEO Sundar Pichai opened the I/O conference that year with a keynote reflecting on the responsibilities of those developing technology, especially AI.

I've come to realise that my red flag law didn't go far enough. We also need to worry about the reverse direction – to prevent people pretending to be using AI when there is no AI in sight. This is such a common deceit that AI researchers have a phrase for it: it's called a 'Wizard of Oz' experiment.

In 2019, for example, an Indian start-up by the name of Engineer.ai claimed it was using AI to automate the development of mobile apps. But it turned out that the apps it built were being written by human engineers. The only AI being used by the company did simple tasks like pricing jobs and assigning humans to tasks. The hype around the start-up's AI no doubt helped the founders raise nearly $30 million.

This is just one of many stories of faking it till you make it. Services like X.ai and Clara have humans pretending to be chatbots scheduling appointments. This is a job so mind-numbingly boring that the human employees doing it were reported to have said that 'they were looking forward to being replaced by bots'.[5] And the business expense management app Expensify had to admit that it had been using humans to transcribe receipts which it had been claiming were processed using its 'smartscan technology'. The receipts were instead posted to Amazon's Mechanical Turk, where low-paid workers were reading and transcribing them – a deceit with very obvious privacy implications.

I'm therefore updating and extending my proposed law. The (extended) Turing Red Flag law states: 'AI should be designed not to be mistaken for a human. Equally, systems should not pretend to be using AI when they are not.'

TECHNICAL FIXES

Our goal, then, should be to build AI systems that we can trust. We don't know how to do this today. But we can at least identify a few of the tools – technical, regulatory and educational – that will help build trust. From a technical perspective, this means building systems that are explainable, auditable, robust, (provably) correct, fair, respectful of privacy, and (where appropriate) transparent.

Take, for example, privacy. An established way to ensure systems are respecting people's privacy is 'privacy by design'. This is a well-established methodology for building systems. It is based on seven foundational principles, such as privacy being the default setting, and the system being preventative rather than remedial. As privacy is not something that can be restored once it has been violated, embedding such a proactive approach into the very design of a system seems a good idea.

This 'by design' meme has taken off for other areas of trustworthy systems. You now hear of building AI systems to be fair by design, explainable by design and, for all I know, 'right up to but not across the creepy line' by design.[6] Trust, like privacy, is difficult to restore once lost, so a proactive approach, built into the architecture of a system, again sounds like a good idea.

As computing moves to the edge, away from the cloud and onto our devices, trust is likely to be easier to achieve. Currently, data leaves your device and ends up in some vast server farm. Once it has left your possession, it's a real challenge to ensure that your privacy is respected, and that nothing untoward happens to your data. It's far better not to give up your data in the first place.

More friction may help. Ironically, the internet was invented to remove frictions – in particular, to make it easier to share data and communicate more quickly and effortlessly. I'm starting to think, however, that this lack of friction is the cause of many problems. Our physical highways have speed and other restrictions. Perhaps the internet highway needs a few more limitations too?

One such problem is described in a famous cartoon: 'On the internet, no one knows you're a dog.' If we introduced instead a friction by insisting on identity checks, then certain issues around anonymity and trust might go away. Similarly, restrictions on how content on social media can be reshared might help prevent the distribution of fake news. And profanity filters might help prevent users posting content that inflames.

On the other side, other parts of the internet might benefit from fewer frictions. Why is it that Facebook can get away with behaving badly with our data? One of the problems here is there's no real alternative. If you've had enough of Facebook's bad behaviour and log off – as I did some years back – then it is you who will suffer most. You can't take all your data, your social network, your posts, your photos to some rival social media service. There is no real competition. Facebook is a walled garden, holding onto your data and setting the rules. We need to open that data up and thereby permit true competition.

REGULATORY FIXES

Technical fixes can only take us so far. It is abundantly clear that we also need more regulation. For far too long the tech industry has been given too many freedoms. Monopolies are starting to form. Bad behaviours are becoming the norm. Many internet businesses are poorly aligned with the public good.

Digital regulation is probably best implemented at the level of nation-states or close-knit trading blocks. In the current climate of nationalism, bodies such as the United Nations and the World Trade Organization are unlikely to reach useful consensus. The common values shared by members of such large transnational bodies are too weak to offer much protection to the consumer.

The European Union has led the way in regulating the tech sector. The *General Data Protection Regulation* (GDPR) and the upcoming *Digital Service Act* (DSA) and *Digital Market Act* (DMA) are good examples of Europe's leadership in this space. A few nation-states have also started to pick up their game. The United Kingdom introduced a Google tax in 2015 to try to make tech companies pay a fair share of tax. And shortly after the terrible shootings in Christchurch, New Zealand, in 2019, the Australian government introduced legislation to fine companies up to 10 per cent of their annual revenue if they fail to take down abhorrent violent material quickly enough. Unsurprisingly, fining tech companies a significant fraction of their global annual revenue appears to get their attention.

It is easy to dismiss laws in Australia as somewhat irrelevant to multinational companies like Google. If they're too irritating, these companies can just pull out of the Australian market. Google's accountants will hardly notice the blip in their worldwide revenue. But national laws often set precedents that get applied elsewhere. Australia followed up with its own Google tax just six months after the United Kingdom. California introduced its own version of the GDPR, the *California Consumer Privacy Act* (CCPA), just a month after the regulation came into effect in Europe. Such knock-on effects are probably the real reason that Google has argued so vocally against Australia's News Media Bargaining Code. They greatly fear the precedent it will set.

Where else might regulation be beneficial? Politics is one area where it is clearly lacking. Most countries have strict laws about the amount of money that can be spent on and the type of political advertising allowed on broadcast media such as television. We don't want the media barons or the people with the most money to win elections on the back of the TV adverts they buy. We want the people with the best ideas and the most democratic support. We recognise the ability of the media to change how people will vote, not necessarily in their own best interests. And yet social media, which is arguably even more persuasive than the old-fashioned media like TV, is much less regulated.

Take micro-targeting, the personalising of adverts based on machine learning and our digital footprints. Is our political debate improved by the ability to micro-target voters with half-truths at very little cost? Freedom of speech is, of course, vitally important. But if you have a message, perhaps you should be limited to broadcasting it for everyone to hear. Would political debate not be more inclusive if the only categories you could target ads with were whether the person was in your constituency and whether they were of voting age?

Some technology companies have seen the writing on the wall. Twitter CEO Jack Dorsey announced on 30 October 2019 that the company would ban all political advertising, as 'political message reach should be earned, not bought'.[7] Other companies seem less enlightened. Facebook continues to insist that it makes little money from political adverts. It's hard, then, to understand why they don't just pull the plug on micro-targeting political adverts. Perhaps we should regulate that they do.

Another area where there is a vital need for more regulation is anti-trust laws. Companies like Google are facing increasing pressure from the US Department of Justice and the European Commission with regard to their monopolistic practices. But current anti-trust regulation seems inadequate to keep on top of their uncompetitive behaviours. Big Tech is getting too big and needs cutting down to size.

In the United States, the *Sherman Antitrust Act* of 1890 focused attention on price fixing. But when the tech giants are giving away many of their services for free, proving that the consumer is paying too much is problematic. Of course, the consumer is ultimately paying inflated prices through the many undesirable externalities in the market – think of Amazon, for example, using its privileged position as both market maker and seller to squeeze other suppliers out of business, or Apple rent-seeking in the premiums it charges software developers to access its app store, or Google depriving traditional print media of income. It is clear that we need a more nuanced view of anti-trust behaviour that goes beyond simply the sticker price on services.

Another aspect of anti-trust regulation in need of attention is the laws governing mergers and acquisitions. In the United States, Section 7 of the *Clayton Antitrust Act* of 1914 seeks to prevent the merger and acquisition of companies which would likely lessen competition substantially. But laws like this are clearly not working.

In 2012 Facebook bought Instagram, a clear competitor in the social media space, for around $1 billion. And in 2014 it bought WhatsApp, again a direct competitor to its messaging service, for $19 billion. Instagram and WhatsApp were doing very well on their own before Facebook bought them. And it was almost certainly Facebook, and not the consumer, who has benefited most from their acquisition.

Google has made a whopping 229 acquisitions in total, spending over $20 billion to acquire market dominance in a number of areas, including advertising, operating systems and smart devices. Amazon has kept a lower profile in the M&A space, acquiring just 87 companies, but has spent much more than Google – over $37 billion by 2020. Amazon's acquisition of competitors, some nascent, as well as companies in adjacent markets, has expanded its market power, again likely to the detriment of the customer.

Moving on to data, new regulations like Europe's GDPR and California's CCPA help to provide the consumer with greater privacy. But these laws are only a start: there are many other emerging issues around data where regulation is needed.

For example, wouldn't it be great if any data about you expired after five years unless you gave explicit consent for it to be retained? Past indiscretions could then be forgotten, as they used to be when I was a boy. Such a rule would barely hinder companies like Amazon. They really only want to know what you might buy today, not what you were interested in decades ago.

Then there are the thorny issues around 'shadow' data. Facebook has profiles on people who have never signed up to or consented to the terms of service. Google is even worse. It maintains a shadow profile on its users containing a vast amount of data and inferences that it has made about you. Thanks to tracking data in Google Maps and Android, it likely knows where you live, where you work, who you associate with, as well as your hobbies and other 'secret interests'. Existing data regulation is largely quiet on the inferences companies make about you, and your rights to correct or delete them.

Finally, there are areas where we urgently need to draw red lines, many of them specific to artificial intelligence. Applications like face recognition and emotion detection simply need to be prohibited from most public and private settings. Similarly, we will need specific laws to deal with tricky issues around autonomy. Lethal autonomous weapons must be banned. And strict limits and liabilities must be introduced in the use of autonomy in other settings, such as in self-driving cars.

Finally, Big Tech will need to be regulated more like Big Pharma. Where are the ethics committees that oversee the use of AI in public settings, like the ethics committees that oversee the use of drugs by the general public? Can we let tech companies run A/B experiments on the public with no oversight when those experiments may change the outcome of elections?

EDUCATIONAL FIXES

Technical and regulatory fixes alone will not be enough. Regulators, politicians and the wider public need to be better informed. For far too long, artificial intelligence has been seen by many as magic. People's perceptions have often been informed more by Hollywood than by reality. I spend a fair amount of my time trying to reset people's concerns, getting them to focus on what is important today, and away from what is science fiction.

I therefore welcome efforts like 'Elements of AI'. As a parting gift at the end of its presidency of the Council of the European Union in December 2019, Finland launched an online crash course in AI for the world. Well, actually for the citizens of Europe, but anyone can sign up for free. I encourage you to do so. I tested it out for you. It's actually rather good![8]

The plan is for the Elements of AI course to be made available in every European Union language. It is currently available in nearly half the official languages: English, Finnish, Swedish, German, Estonian, Latvian, Lithuanian, French, Dutch and Maltese. The course has enrolled around half a million students from over 100 countries. Pleasingly, some 40 per cent of students are female. In Nordic countries, women make up nearly 60 per cent of participants.

Initiatives like Elements of AI are essential to democratise artificial intelligence. If AI remains magic, then many of us will surely be taken advantage of. People will struggle to make informed choices about where to let AI into their lives, and where to lock it out. Understanding what AI is, how it works and what its limitations are is essential to building trust.

It amazes me that people are surprised to discover that smart speakers like Amazon's Alexa are always listening. How do you think they can wake up when we say 'Alexa'? George Orwell predicted that a totalitarian state might forcibly put listening devices into everyone's home that you couldn't turn off. He could never have imagined that people would do so freely, and even pay for the product!

An informed population would be more equipped to hold the tech companies to account. Consumers would more likely walk away when such companies behaved badly. People would make better choices about using AI, about where to trust and where to mistrust the technology. There are no downsides that I can see to having a more informed citizenry.

Take fake news. If more people understood the limits of current technologies to process natural language, they would treat Mark Zuckerberg's claim to Congress that Facebook will use AI to detect fake news with the scepticism it deserves.[9] They would appreciate that it might be several decades before AI can understand natural language with sufficient nuance to spot the difference between real and fake news. Imagine how toxic social media could be by then.

THE GIFT OF THE MACHINES

Hopefully by now you understand a little more about the complex issues surrounding AI and ethics. It is worth remembering what amazing machines we are going to build, even if their moral decision-making is going to be somewhat challenged.

I used to think the greatest gift these machines would give us is time. After all, time is our most valuable commodity. Each of us is given just three score years and ten on this Earth to do with as best we decide. Not even the richest among us can buy much more time than this.[10] However, artificial intelligence can give us back some of that time. In fact, AI can usefully liberate us from the four Ds: the dirty, the dull, the difficult and the dangerous.

Luigi, a robot developed at MIT, is performing the *dirty* job of sampling sewage to test drug use and disease prevalence in urban populations. RIPPA, a weeding robot developed at the Australian Centre for Field Robotics, is taking over a job that most of us would find immensely *dull.* AILIRA (the Artificially Intelligent Legal Information Research Assistant), developed by Cartland Law in Adelaide, can answer *difficult* questions about tax law. (In many people's eyes, that probably counts as a *dull* job too.) And my own lab, Data61, has developed a fully autonomous mine-clearing robot. This surely counts as a *dangerous* job that machines have taken over.

Getting machines to do jobs that are dirty, dull, difficult or dangerous should not trouble us. When someone tells me that a particular job has been automated, I usually think that we should rejoice. The fact that we can get a machine to do this job says that this was likely a repetitive and boring job. And we should probably never have got a human to do this job in the first place.

Of course, replacing a human with a machine is not without other challenges. What happens to the person who is replaced? Do they have a rewarding job to move on to? Are the financial benefits of automation being shared? And who is accountable for the machine when things go wrong?

So if time isn't the greatest gift that machines will give us, what is? In the long term, I suspect it will be a greater appreciation, perhaps even an enhancement, of our own humanity. Intelligent machines could actually make us better humans. There are at least four ways in which this can happen.

The first way that intelligent machines might make us better humans is by rendering human relationships even more valuable. Machines today have limited emotional and social intelligence. Even if we can program machines in the future with greater emotional and social intelligence, I doubt we will empathise with them as we do with humans. A machine won't fall in love, mourn a dead friend, stub a toe and jump around in pain, enjoy the delicate smell of tuber roses, roll around the floor in laughter or be swept along by a performance of classical music. These are *uniquely* human experiences. And since machines won't share these experiences with us, we will never relate to them as we do to each other. We are intensely social animals. It is this society, along with technology, that has allowed our civilisations to advance. And society is built upon relationships. We will never have as rich relationships with machines.

The second way that intelligent machines could make us better humans is by making human creation more valuable. Machines will create many of life's necessities, so their price will plummet. On the other hand, things made by the human hand will necessarily be rarer. And, as economists have taught us, such handmade objects will be reassuringly expensive. We already see this today in hipster culture. There is an appreciation of the handmade, the artisanal and the artistic. We see a new appetite for artisanal bread, homemade cheese, craft beer, bespoke suits, custom-made shoes, live concerts and fine art. In some artificially intelligent future, machines may be able to compose music to rival that of Johann Sebastian Bach, to write sonnets to equal William Shakespeare's, to paint as well as Pablo Picasso or to pen novels that compare with the great works of Jane Austen. But will we care? Such AI will not speak to those uniquely human experiences that much great art addresses – love, loss and mortality.

The third way that intelligent machines could make us better humans is by being *artificially* intelligent. The early indications are that AI will be a very different form of intelligence than our own *natural* human intelligence. We already have plenty of examples of the differences. On the one hand, AI can surpass human intelligence by finding insights in data sets too large for humans to comprehend. On the other hand, AI can be immensely brittle compared to human intelligence. It can fail on tasks that even a four-year-old could manage. If recent progress is an indicator, artificial intelligence is likely to be a much more logical, statistical and unemotional type of intelligence compared to human intelligence. Think Spock rather than Captain Kirk. Such a future may help us appreciate more our own messy human intelligence.

The fourth and final way that intelligent machines could make us better humans is by instilling in us a greater appreciation for, and perhaps an enhancement of, human values. Indeed, AI could lead to a golden age for philosophy, an age in which we can live better lives. The goal of trying (and in many cases failing) to program machines with ethical values may lead us to a deeper understanding of our own very human values. It will force us to answer questions that society has often dodged in the past. How do we value different lives? What does it mean to be fair and just? In what sort of society do we want to live?

Our future is not infallible super-intelligence and immortality. That is the future of our machines. Our future is the opposite, ever fallible and mortal.

Let us therefore embrace what makes us human. It is all that we will ever have!

THE CHILD OF OUR BRAINS

31 DECEMBER 2061

What seems a lifetime ago, I wrote a trilogy of books about the world that technologies like artificial intelligence were starting to make. And now Big Ben is about to herald the start of 2062, and we are all living in that AI-enabled future. It's a good moment, then, to look back on those 40 years, to consider all the predictions I got wrong, and to marvel at how far the human race has come since the eventful year of 2020.

I argued then that we needed a global shock to precipitate change, to reform our society, to reduce inequality and to live life more sustainably. And that technologies like artificial intelligence could provide us with some of the tools to make the planet a better place.

It's hard to know which of the many events of 2020 was the most important shock. Was it the bushfires in Australia and California, which signalled that the climate emergency could no longer be ignored? Or was it the rejection of Trump and his divisive politics, which began the shift to a more compassionate society? Or was it something much smaller, the SARS-CoV-2 virus (just 125 nanometres in diameter), which kickstarted those two decades of fundamental reform across the planet?

In 2021, as the world started to vaccinate against COVID-19, people around the globe were awakening to the idea that change was possible. Decades of neoliberalist politics had locked us into policies that tried to balance budgets, reduce taxes and seek economic growth at any environmental cost. While these policies did grow the economy, they did so at the expense of the planet. And at the expense of equality within society. The planet was literally burning while the rich were getting richer.

The COVID-19 pandemic showed that alternative, kinder, more equitable and sustainable paths were possible. Leaders like President Donald Trump of the United States were exposed for who they really were. And compassionate and inclusive leaders – most of them female, it has to be noted – such as Prime Minister Jacinda Ardern of New Zealand, led the way in showing that we could build a kinder society to improve the wellbeing of everyone, not just the richest 1 per cent.

It's hard to remember how sweet freedom tasted back then, after repeated cycles of lockdown. As we cautiously emerged from our isolated social bubbles, solidarity between the citizens of the planet took hold. We were in this together. Every life mattered.

There was a lot that needed to be changed. The mass demonstrations against racism were only the start. Climate. Poverty. Inequality. Sexism. Democracy. All these movements were fired up. Change was possible. And change began to happen. Of course, there were setbacks along the way. But life never did go back to normal.

Globalisation was reset. We didn't throw away the benefits of international trade and cooperation, but the pandemic awakened an appreciation for simpler, more local and, in many ways, more 'old-fashioned' values. We stopped flying strawberries around the planet, but we didn't stop sending medicine to the developing world, or exporting our scientific advances for everyone to share.

The economic pressures of the pandemic hastened automation and the adoption of technologies like artificial intelligence. The 2020s saw a decade-long fight to get the Big Tech giants to pay more taxes, to moderate fake news and to behave more responsibly within the markets in which they acted.

The 2020s also saw a decade-long fight to rein in those governments that had tried to use the pandemic as an excuse to control their citizens more. It seemed that *Nineteen Eighty-Four* became 2024. But the surveillance state effectively ended with the Great Internet Blackout, which cyber-activists unleashed on us all that fateful New Year's Eve in 2026.

As the machines began to take over more jobs, there was a flowering of the arts, crafts and community. It was the start of a second renaissance. A rebirth of the human race.

That's enough from me – my time is coming to an end. What a privilege it was to live through such change. To see the young grasp the wicked problems we all faced back then. And to see them use technologies like AI to build a better and more humane future.

ABOUT THE AUTHOR

Ever since he was young, Toby Walsh has been dreaming about building machines that might think. He was named by the *Australian* newspaper as one of the 'rock stars' of Australia's digital revolution. His family think the rock star part is highly improbable.

He is a Laureate Fellow and Scientia Professor of Artificial Intelligence at UNSW and at CSIRO Data61. He is a Fellow of the Australian Academy of Science, of the Association for Computing Machinery, of the American Association for the Advancement of Science, and of the Association for the Advancement of Artificial Intelligence. He has held research positions in Australia, England, France, Germany, Ireland, Italy, Scotland and Sweden.

He appears regularly on TV and radio, talking about the impact of AI and robotics. He also writes frequently for newspapers and magazines, such as the *Guardian, New Scientist, American Scientist* and *Cosmos*. His Twitter account, @TobyWalsh, has been voted one of the top ten to follow to keep abreast of developments in AI.

Black Inc. published his two earlier books on AI, each written for a general audience. The first, *It's Alive! Artificial Intelligence from the Logic Piano to Killer Robots*, looked at the past, present and near future of AI. The book also appeared in the United Kingdom under the title *Android Dreams: The Past, Present and Future of AI*, and in the United States under the title *Machines that Think: The Future of Artificial Intelligence*. His second book, *2062: The World that AI Made*, considered our more distant future, when machines might match or even exceed human intelligence. The two books are available in Arabic, Chinese, English, German, Korean, Polish, Romanian, Russian, Turkish and Vietnamese.

Above all, Toby is passionate about placing limits on AI to ensure it improves the quality of all our lives. He has spoken at the United Nations in both New York and Geneva about the need to ban lethal autonomous weapons (aka 'killer robots'). You can read more at his blog, thefutureofai.blogspot.com.

ACKNOWLEDGEMENTS

I want to thank my agent, Margaret Gee, who helped put this book into your hands, along with my editors, Julian Welch and Kate Morgan, at Black Inc. for doing an excellent job getting this into a final form. Any errors that remain are my own.

There are many other people I want to thank:

My parents, who supported my AI dreams from an early age.

My twin brother, who has always helped keep my feet on the ground.

My academic colleagues at UNSW Sydney, CSIRO Data61 and elsewhere, especially my research collaborators and students, who provide the stimulating environment in which I continue to explore these dreams.

Nadia Laurinci and her team from Laurinci Speakers, for managing all my speaking engagements.

But above all I want to thank my family, who generously gave me the time to write a *third* book. It's an enjoyable itch.

NOTES

AI

1   Neil Postman (1992) *Technopoly: The Surrender of Culture to Technology*, New York, Knopf.

2   See tinyurl.com/PostmanFiveLessons.

3   See tinyurl.com/Simulmatics.

4   Norbert Wiener (1950) *The Human Use of Human Beings: Cybernetics and Society*, New York, Houghton Mifflin.

5   To watch the BBC video where this close relationship between Facebook and Cambridge Analytica is revealed, see tinyurl.com/ FacebookLiedTheyKnew.

6   See tinyurl.com/ZuckerbergCongress.

7   See tinyurl.com/FacebookSwingsElections.

The People

1   See tinyurl.com/WorldAIPhDs.

2   One year, I asked my travel agent to book me on a research trip from my home in Cork, to a conference in Cuba, via a meeting in Cyprus, and on to a summer school in Canberra, before returning home to Cork. My travel agent joked I was just picking places from the atlas that begin with the letter C.

3   Antarctica's emperor penguin population has been counted using machine-learning tools trained on images from Oxford University's Penguin Watch

Project. And while I don't believe any AI conference has been held in
Antarctica (yet), a technology conference called AntarctiConf 2020 was
held on board a 16-day cruise to Antarctica in January 2020, just before
the COVID-19 pandemic closed the world down.

4   Actually $2^{82,589,933}-1$ is the *largest known* prime number. It would have
24,862,048 digits when written out in longhand, but this would, I
imagine, make for a very boring book. It is also not *any old* prime number,
but a very special type of prime number. It is a Mersenne prime number,
since it is one less than a power of two. Such prime numbers are named
after Marin Mersenne, a French friar who studied them in the early
seventeenth century. While this is the largest prime number to have been
discovered by mathematicians, there is no *largest* prime number. Over
2000 years ago, Euclid proved that, while prime numbers become rarer
and rarer as numbers get bigger, prime numbers never end. And while we
don't know if $2^{82,589,933}-1$ is the largest Mersenne prime or not, we do
know, thanks to the law of the excluded middle, that one of these
statements must be true: either it is the largest Mersenne prime, or there
is a larger such prime.

5   In the interests of full disclosure, I am part of the problem, being one of
those white males.

6   Peter E. Hart, Niles J. Nilsson & Bertram Raphael (1968) 'A Formal
Basis for the Heuristic Determination of Minimum Cost Paths', *IEEE
Transactions of Systems Science and Cybernetics*, vol. 4, no. 2, pp. 100–07.

7   See tinyurl.com/TimeMagazine100.

8   David E. Rumelhart, Geoffrey E. Hinton & Ronald J. Williams (1986)
'Learning Representations by Back-propagating Errors', *Nature*, 323,
pp. 533–36.

9   In 2012, Andrew Ng, Jeff Dean and colleagues at Google's secretive X
Laboratory built one of the largest neural networks up to that time. They
needed 16,000 cores across 1000 computers to train the network's billion
weights. And after three days of training on still images from random
YouTube videos, the network famously taught itself to recognise cats.
What else would you expect to come out of YouTube? See Q.V. Le et al.
(2012) 'Building High-level Features Using Large Scale Unsupervised
Learning', in *Proceedings of the 29th International Conference on Machine
Learning* (ICML2012), Madison, WI, Omnipress, pp. 507–14.

10 While Grace Hopper was working on a Mark II Computer at Harvard University in 1947, an error was uncovered, caused by a moth stuck in a relay. This incident is why programmers still talk about finding and removing 'bugs' from their programs.

11 If you want an alternative, in my first book, *It's Alive! Artificial Intelligence from the Logic Piano to Killer Robots*, I suggested we call Bengio, Hinton and LeCun the 'three musketeers' of deep learning.

12 In *It's Alive!*, I told the story of how I nearly drowned John McCarthy in Sydney Harbour, and how, as a result, I feared I would end up a sad footnote in a history of AI. This footnote is another attempt to do just that.

13 All dollar amounts are in US dollars.

14 In April 2020, SoftBank announced its first annual loss of over $16 billion in its $100-billion Vision Fund. The loss was in part due to investments in companies like Wag, a dog-walking service, and Zume, a robot pizza company.

15 For more on Rand, see Chris Matthew Sciabarra (1995) *Ayn Rand: The Russian Radical*, University Park, Pennsylvania State University Press.

16 You can read the *Declaration of the Independence of Cyberspace* in full at eff.org/cyberspace-independence.

The Companies

1 Nick Bostrom (2014) *Superintelligence: Paths, Dangers, Strategies*, Oxford, Oxford University Press.

2 In 2018, Facebook agreed with an independent report that found it had failed to prevent its platform from being used to 'incite offline violence' among its 18 million users in Myanmar. Thousands of people died in the civil unrest in Myanmar, and over 700,000 Rohingya were forced to flee for their lives to neighbouring Bangladesh.

3 See tinyurl.com/KendallTestimony.

4 Julia Angwin & Terry Parris Jr, 'Facebook Lets Advertisers Exclude Users by Race', *ProPublica*, 28 October 2016, propublica.org/article/facebook-lets-advertisers-exclude-users-by-race.

5   'There are only two industries that refer to their customers as "users": illegal drugs and software,' said Edward Tufte, American statistician and professor emeritus at Yale.

6   Julia Angwin, Ariana Tobin & Madeleine Varner, 'Facebook (Still) Letting Housing Advertisers Exclude Users by Race', *ProPublica*, 21 November 2017, propublica.org/article/facebook-advertising-discrimination-housing-race-sex-national-origin.

7   Jessica Livingstone (2007) *Founders at Work: Stories of Startups' Early Days*, New York, Springer-Verlag.

8   See https://ai.google/principles.

9   See ibm.com/blogs/policy/trust-principles.

10 Of course, shortly after I wrote this chapter, Lemonade appeared to have committed a serious ethical blunder. They are currently facing a class-action lawsuit for using biometric data without consent with facial-recognition software that analyses videos submitted in the claims process.

Autonomy

1   My colleague Professor Genevieve Bell of ANU has whimsically asked: 'Who is the self driving a self-driving car?'

2   There are many flavours of Uber: UberX, UberPool, UberEats … By 2050, all of them will be autonomous except for one, UberChauffeur. This will be a premium service where you pay extra for a person who wears a smart uniform with a cap to sit in the driver's seat, open the car door and carry your luggage to the kerb. They won't drive; the Uber vehicle will still be self-driving. But you will pay extra, as in other parts of your life, for personal service that connects you with a real human being.

3   See tinyurl.com/aitrends2017.

4   When I started writing this chapter, Tesla had a market cap of just $400 billion. By the time I had finished the book, this had nearly doubled. I suspect that, by the time you read this, it may have doubled again.

5   While Level 5 is the highest level of autonomy in the Society of Automotive Engineers' classification scheme, there are actually six levels in total. Level 0 is, however, no automation at all.

6　See tinyurl.com/ElonMuskLidar.

7　See tinyurl.com/GermanEthicsCode.

8　I don't mean they literally paint the road green with a paintbrush. But if you watch a demonstration of a self-driving car on YouTube, you will often see that the video image of the open road in front of the car is shaded green.

9　Edmond Awad et al. (2020) 'Crowdsourcing Moral Machines', *Communications of the ACM*, vol. 63, no. 3, pp. 48–55.

10　Edmond Awad et al. (2020) 'Universals and Variations in Moral Decisions Made in 42 Countries by 70,000 Participants', *Proceedings of the National Academy of Sciences*, vol. 117, no. 5, pp. 2332–37.

11　See tinyurl.com/KillerRobotPoll.

12　Poseidon is a troubling name for an autonomous nuclear weapon. Poseidon is the god of the sea, earthquakes, storms and horses. He is one of the most bad-tempered, moody and greedy Olympian gods.

13　See tinyurl.com/DronePapers.

14　If you agree with me and the thousands of other AI researchers who have signed the open letter to the United Nations, you can sign it too: visit https://futureoflife.org/open-letter-autonomous-weapons.

15　An experimental study of facial-recognition software from 16 major commercial suppliers in 2017 by the National Institute of Standards and Technology in the United States found false positive rates of over 90 per cent in some circumstances; see tinyurl.com/NISTReport2017.

16　See tinyurl.com/DoDEthicalPrinciples.

Humans v. Machines

1　A petabyte is 10^{15} bytes of information. See tinyurl.com/BrainVInternet.

2　Geoffrey Jefferson (1949) 'The Mind of Mechanical Man', *British Medical Journal*, vol. 1, p. 1105.

3　See Daniel Kahneman (2011) *Thinking, Fast and Slow*, New York, Farrar, Straus & Giroux.

4   Article 13 of the Universal Declaration of Human Rights states:

(1) Everyone has the right to freedom of movement and residence within the borders of each state.

(2) Everyone has the right to leave any country, including his own, and to return to his country.

This universal right to freedom of movement has been somewhat ignored during the COVID-19 pandemic.

5   Boston Dynamic's Atlas robot was also able to open a door using the door handle, putting to rest the idea that you only had to hide beyond a closed door to escape a marauding robot.

6   See tinyurl.com/CNNrobotdog.

7   See stoprobotabuse.com.

8   See tinyurl.com/EURobotRights.

9   If you want to be impressed, check out the video on YouTube of the Guinness World Record for solving the Rubik's cube: see tinyurl.com/fastestcube.

10   Gerald Tesauro et al. (2013) 'Analysis of Watson's Strategies for Playing Jeopardy!', *Journal of Artificial Intelligence Research*, vol. 47, pp. 205–51.

Ethical Rules

1   Howard Rheingold (1985) *Tools for Thought*, Englewood Cliff, New Jersey, Prentice Hall.

2   I.J. Good (1965) 'Speculations Concerning the First Ultraintelligent Machine', *Advances in Computers*, vol. 6, pp. 31–88.

3   See tinyurl.com/goodsrule.

4   Seneca the Younger, 'XLVII: On Master and Slave', *Epistulae Morales ad Lucilium* (*Moral Letters to Lucilius*).

5   Asimov's three laws were first published in the 1942 short story 'Runaround', which was later included in the 1950 collection *I, Robot*.

6   See tinyurl.com/asimov3laws.

7   See tinyurl.com/duckmeetscar.

8   See tinyurl.com/newrobotlaws.

9   Robin R. Murphy & David D. Woods (2009) 'Beyond Asimov: The Three Laws of Responsible Robotics', *IEEE Intelligent Systems*, vol. 24, no. 4, pp. 14–20.

10 See tinyurl.com/RobotsDoNoHarm.

11 See https://futureoflife.org/ai-principles.

12 The precautionary principle has been used in numerous international laws and rules. It is, for example, the 15th principle of the United Nations' 1992 Rio Declaration on Environment and Development.

13 See tinyurl.com/euethics.

14 It would be unnecessarily rude to California and excessively generous to the European Union to recall that Oscar Wilde actually argued: 'Imitation is the sincerest form of flattery that mediocrity can pay to greatness.'

15 Apple potentially faces an even larger fine of around $15 billion for its legendary 'double Irish' tax avoidance strategy. In August 2016 the European Commission ruled that Apple benefited from illegal tax arrangements from 2003 to 2014. Not surprisingly, Apple appealed the decision. More surprisingly, the Irish government also appealed, telling the European Commission that it didn't want to receive the $15 billion fine. Presumably, the Irish government preferred to keep multinational companies sweet with tax breaks rather than pay off around 5 per cent of Ireland's national debt. The General Court of the European Union reversed the decision against Apple in July 2020. However, the European Commission has announced it will appeal this reversal. It looks like this legal saga still has a long way to run.

16 Why are flight recorders called black boxes when they are bright orange?

17 See tinyurl.com/ChinaAIPrinciples.

18 Mark Twain, 'The Work of Gutenberg', *The Hartford Daily Courant*, 27 June 1900, p. 7.

19 For a more detailed discussion, see Luciano Floridi & Josh Cowls (2019) 'A Unified Framework of Five Principles for AI in Society', *Harvard Data Science Review*, vol. 1.

20 See tinyurl.com/7ethicaltraps.

21 See tinyurl.com/DennettWired.

Fairness

1   See tinyurl.com/TIMEin2011.

2   See tinyurl.com/NSWpolicetargets.

3   'COMPAS' stands for Correctional Offender Management Profiling for Alternative Sanctions.

4   See tinyurl.com/ProPublicaCOMPAS.

5   Of course, there is an argument for the reverse, that crime *causes* broken homes, as crime often results in jail, and jail time often results in parents being separated from their children. But the COMPAS questionnaire presumes the reverse causal link: that broken homes cause crime.

6   Julia Dressel & Hany Farid (2018) 'The Accuracy, Fairness, and Limits of Predicting Recidivism', *Science Advances*, vol. 4, no. 1, DOI: 10.1126/sciadv.aao5580.

7   A linear classifier just tries to find simple linear relations between features and classes. For instance, it might decide that a person is high-risk if their number of priors less one-tenth of their age in years is greater than zero.

8   Sahil Verma & Julia Rubin (2018) 'Fairness Definitions Explained', in *FairWare '18: Proceedings of the International Workshop on Software Fairness*, Association for Computing Machinery, New York, pp. 1–7.

9   Jon M. Kleinberg, Sendhil Mullainathan & Manish Raghavan (2017) 'Inherent Trade-Offs in the Fair Determination of Risk Scores', in *8th Innovations in Theoretical Computer Science Conference*, ITCS 2017, 9–11 January 2017, Berkeley, pp. 43:1–43:23.

10 False positives are also called 'Type 1 errors', while false negatives are called 'Type 2 errors'.

11 Stephen Hawking was warned that each equation in *A Brief History of Time* would halve the sales of his book. I'm hoping the same isn't true for fractions.

12 See tinyurl.com/PartnershipOnAIReport.

13 Ronald C. Arkin (2010) 'The Case for Ethical Autonomy in Unmanned Systems', *Journal of Military Ethics*, vol. 9, no. 4, pp. 332–41.

14 Allison Koenecke et al. (2020) 'Racial Disparities in Automated Speech Recognition', *Proceedings of the National Academy of Sciences*, vol. 117, no. 14, pp. 7684–89.

15 The 'word error rate' is the percentage of words that needed to be added, deleted or changed to give the correct transcription of the speech.

16 See tinyurl.com/FacialRecognitionArrest.

17 S. Ahmed et al. (2021) 'Examining the Potential Impact of Race Multiplier Utilization in Estimated Glomerular Filtration Rate Calculation on African-American Care Outcomes', *Journal of General Internal Medicine*, vol. 36, pp. 464–71.

18 Z. Obermeyer et al. (2019) 'Dissecting Racial Bias in an Algorithm Used to Manage the Health of Populations', *Science*, vol. 366, no. 6464, pp. 447–53.

19 Michael J. Rosenfeld, Reuben J. Thomas & Sonia Hausen (2019) 'Disintermediating Your Friends: How Online Dating in the United States Displaces Other Ways of Meeting', *Proceedings of the National Academy of Sciences*, vol. 116, no. 36, pp. 17753–58.

20 Uber does not classify its drivers as employees but as independent contractors. However, like the State of California, I don't buy this. It appears to be largely an excuse to give them fewer rights and worse pay. This is not something a decent society should permit.

21 See tinyurl.com/AmazonAITool.

22 Richard Landers (2012) 'Unfolding the IKEA Effect: Why We Love the Things We Build', *Journal of Consumer Psychology*, vol. 22, no. 3, pp. 453–60.

23 Alvin E. Roth & Elliot Peranson (1997) 'The Effects of the Change in the NRMP Matching Algorithm', *Journal of the American Medical Association*, vol. 278, no. 9, pp. 729–32.

Privacy

1 'Skynet' is the name of the fictional and conscious AI system threatening the future of humanity in the *Terminator* films.

2 See tinyurl.com/FaceInRockConcert.

3   Vint Cerf has been called one of the 'fathers of the internet', but I'm trying to avoid that sexist pitfall here. I had the pleasure of interviewing him on stage once, so I can attest to the fact that he is indeed a nice guy. You can watch the interview and learn more about his many contributions to the technology in your life at tinyurl.com/VintCerfVideo.

4   Samuel D. Warren & Louis D. Brandeis (1890) 'The Right to Privacy', *Harvard Law Review*, vol. 4, no. 5, pp. 193–220.

5   Alan F. Westin (1967) *Privacy and Freedom*, New York, Atheneum, p. 487.

6   See tinyurl.com/BrianActon.

7   This is, of course, an allusion to a famous sequence of instant messages that Mark Zuckerberg sent while still a student at Harvard.

    Zuck: yea so if you ever need info about anyone at harvard

    Zuck: just ask

    Zuck: i have over 4000 emails, pictures, addresses, sns

    Friend: what!? how'd you manage that one?

    Zuck: people just submitted it

    Zuck: i don't know why

    Zuck: they 'trust me'

    Zuck: dumb f—ks

8   See tinyurl.com/IndiaFaceRecognition.

9   See tinyurl.com/IndiaFaceRecognition2.

10 See tinyurl.com/IndiaFaceRecognition3.

11 See tinyurl.com/FaceRecognitionArrest.

12 See tinyurl.com/FaceRecognitionFine.

13 See tinyurl.com/MerkleyStatement.

14 Yilun Wang & Michal Kosinski (2018) 'Deep Neural Networks Are More Accurate than Humans at Detecting Sexual Orientation from Facial Images', *Journal of Personality and Social Psychology*, vol. 114, no. 2, pp. 246–57.

15 Michal Kosinski (2021) 'Facial Recognition Technology Can Expose Political Orientation from Naturalistic Facial Images', *Nature Scientific Reports*, vol. 11, no. 100.

16 See tinyurl.com/CreepySchmidt.

17 See tinyurl.com/GmailPrivacyCase.

18 See tinyurl.com/GmailAppDeveloper.

19 Aleecia M. McDonald & Lorrie Faith Cranor (2008) 'The Cost of Reading Privacy Policies', *I/S: A Journal of Law and Policy for the Information Society*, vol. 4, no. 3, pp. 543–68.

The Planet

1 The online version of this interview was corrected after I complained about this quote, but you can read the original erroneous quote I reproduce here on the Factiva database: 'Australia Joins World-first AI Group to Tackle Ethics, Commercialisation', *Australian Financial Review*, 15 June 2020.

2 Anders S.G. Andrae & Tomas Edler (2015) 'On Global Electricity Usage of Communication Technology: Trends to 2030', *Challenges*, vol. 6, pp. 117–57.

3 Eric Masanet et al. (2020) 'Recalibrating Global Data Center Energy-use Estimates', *Science*, vol. 367, no. 6481, pp. 984–86.

4 Lasse F. Wolff Anthony, Benjamin Kanding & Raghavendra Selvan (2020) 'Carbontracker: Tracking and Predicting the Carbon Footprint of Training Deep Learning Models', ICML Workshop on Challenges in Deploying and Monitoring Machine Learning Systems.

5 Microsoft Azure provides the compute infrastructure to OpenAI as part of Microsoft's $1-billion investment in OpenAI. In return, Microsoft has been given an exclusive licence to GPT-3.

6 See tinyurl.com/AIIndex2019.

7 Danny Hernandez & Tom B. Brown (2020) 'Measuring the Algorithmic Efficiency of Neural Networks', arXiv report 2005.04305.

8 See tinyurl.com/AICement.

9   R. Vinuesa et al. (2020) 'The Role of Artificial Intelligence in Achieving the Sustainable Development Goals', *Nature Communications*, vol. 11, no. 1, p. 233.

The Way Ahead

1   See tinyurl.com/EUAIGuidelines.

2   See transparency.facebook.com.

3   See tinyurl.com/FacebookNoTrust.

4   See tinyurl.com/TuringRedFlag.

5   See tinyurl.com/BoringJobs.

6   With credit to Eric Schmidt, ex-CEO and chairman of Google: see tinyurl.com/CreepySchmidt.

7   See tinyurl.com/TwitterMicrotargetBan.

8   You can sign up for 'Elements of AI' at elementsofai.com.

9   See tinyurl.com/AI4FakeNews.

10   For the nitpicker, it is true that life expectancy in some richer countries has increased beyond three score years and ten to over four score years. And richer people within those countries tend to live a little bit longer still, due to better diet and so on. But the number of extra years alive that money can buy is still rather limited. Paul Allen's immense wealth from co-founding Microsoft could not save him from dying from the complications of Non-Hodgkin lymphoma at just 65 years of age. Nor could the considerable wealth of Steve Jobs, co-founder of Apple, help him defeat pancreatic cancer, which killed him at just 56 years of age.

INDEX

ᵗ